INTERNATIONAL SERIES OF MONOGRAPHS
IN ANALYTICAL CHEMISTRY

General Editors
R. BELCHER AND H. FREISER

VOLUME 54

ORGANIC REAGENTS IN METAL ANALYSIS

ORGANIC REAGENTS IN METAL ANALYSIS

by

K. BURGER D.Sc.

Professor of Inorganic and Analytical Chemistry
at the L. Eötvös University, Budapest, Hungary

PERGAMON PRESS

Oxford • New York • Toronto
Sydney • Braunschweig

Pergamon Press Ltd., Headington Hill Hall, Oxford
Pergamon Press Inc., Maxwell House, Fairview Park, Elmsford, New York 10523
Pergamon of Canada Ltd., 207 Queen's Quay West, Toronto 1
Pergamon Press (Aust.) Pty. Ltd., 19a Boundary Street,
Rushcutters Bay, N.S.W. 2011, Australia
Vieweg & Sohn GmbH, Burgplatz 1, Braunschweig

Copyright © 1973 Akadémiai Kiadó, Budapest

All rights reserved. No part of this publication may be reproduced, stored in a retrieval system, or transmitted, in any form or by any means, electronic, mechanical, photocopying, recording or otherwise, without the prior permission of Pergamon Press Ltd.

First edition 1973

This book is the enlarged English version of

"Szerves reagensek a fémanalízisben"
published by Műszaki Könyvkiadó, Budapest

Translated by

T. Mohácsy

Joint edition published by Pergamon Press Ltd., and
Akadémiai Kiadó, Publishing House of the Hungarian Academy of Sciences

Library of Congress Cataloging in Publication Data

Burger, Kálmán.
  Organic reagents in metal analysis.

  (International series of monographs in analytical chemistry, v. 54)
  "Enlarged English version of Szerves reagensek a fémanalízisben ... Translated by T. Mohácsy."
  Bibliography: p.
  1. Metals--Analysis. 2. Chemical tests and reagents. 3. Complex compounds. I. Title.
QD133.B913 1973   546'.3   72-10479
ISBN 0-08-016929-5

Printed in Hungary

CONTENTS

Preface   9

Introduction   11

CHAPTER 1

The Analytical Selectivity of Reactions Based on Complex Formation   15

I. *Effect of the stabilities of complexes upon selectivity*   15

    1. Factors determining the stabilities of complexes   15
        Effect of the metal ion   15
        Effect of the ligand   18
        Steric effects   20
    2. The ratio of stepwise stability constants and analytical selectivity   23
    3. The significance of stability constants and equilibrium calculations in
        determining the analytical selectivity   26
        The conditional equilibrium constants and their application   27
        Calculation of conditional equilibrium constants   29
        Consideration of the formation of protonated and hydroxo mixed complexes   34
        Masking and demasking   35
        Selectivity index   38

II. *Effect of changes in the electronic structure of the central atom upon the selectivity of the reaction*   41

    1. Ligand-field effect   42
    2. Effect of back-coordination   45

III. *Analytical selectivity of the formation of mixed complexes*   47

    Conditions for the analytically selective formation of mixed complexes   47
    Factors determining the solubilities of complexes   49
    Absorption of complexes   51

IV. *Functional groups and analytical selectivity*   54

    8-Hydroxyquinoline (oxine) and its derivatives   56
    4-Hydroxybenzothiazole   56
    Nitrosonaphthol ligands   57
    2,2′-Dipyridyl and 1,10-phenanthroline   58
    Dioximes   59
    α-Hydroxyaldoximes   62
    Pyrogallol   63

β- and o-Amino acids	64
α-Hydroxycarboxylic acids	65
Rhodanine and its derivatives	65
Organic nitro compounds	66

CHAPTER 2

Application of Selective Organic Reagents in Quantitative Chemical Analysis — 67

I. *The methods of application of selective organic reagents* — 67

1. Organic reagents in gravimetric analysis	68
Precipitation from homogeneous solution	69
Enrichment by co-precipitation	70
2. Organic reagents in volumetric analysis	70
3. Organic reagents in spectrophotometric analysis	72
4. Organic reagents in polarography	75
5. Organic reagents in chromatography and ion-exchange	78

II. *Some important organic metal reagents* — 79

1. Phenanthroline and related compounds	80
1,10-Phenanthroline	80
Neocuproine	81
2,2′-Dipyridyl	82
2. Oximes	83
Dimethylglyoxime	83
Nioxime	88
α-Furyldioxime	88
Salicylaldoxime	89
α-Benzoinoxime	90
Daxime	91
3. Nitroso compounds	92
α-Nitroso-β-naphthol	92
β-Nitroso-α-naphthol	93
Cupferron	94
4. Nitro compounds	94
α-Nitro-β-naphthol	94
Picrolonic acid	95
5. Oxine and its derivatives	96
8-Hydroxyquinoline (Oxine)	96
8-Hydroxyquinaldine	105
6. PAN	105
7. Organic acids	108
Anthranilic acid	108
Quinaldic acid	109
Mandelic acid	110
8. Flavones	111
Morin	111
Quercetin	112
9. β-Diketones	112
Acetylacetone	112
Thenoyltrifluoroacetone (TTA)	116
10. Dithizone	118
11. Dithiocarbamates	124
Sodium diethyldithiocarbamate	124
Diethylammonium diethyldithiocarbamate	128

12. Further sulfur-containing reagents	129
Rubeanic acid	129
Thionalide	130
Thioglycollic acid anilide	131
Thioacetamide	132
13. Arsonic acids	132
Phenylarsonic acid	132
Arsanilic acid	133
Arsenazo I (Uranone)	134
Arsenazo III	134
14. Kalignost	137

CHAPTER 3

Summarizing Tables	139
Some more important organic reagents and their scope of application	139
Spectrophotometric determination of metals	185
Thermal stabilities and gravimetric factors of some of the more important metal complexes utilized in gravimetric analysis	205
Stability constants of the chelates of complexing agents used as analytical reagents	207
Signs and abbreviations used	210
Some more important organic solvents	233
References	239
Author Index	251
Subject Index	259

PREFACE

THE tremendous scientific and technical development during the past twenty years has led to increased demands with regard to the production and quality control of highly pure materials. Research continues to uncover the increasingly important role of trace contaminants in more and more fields, ranging from industrial technology to living organisms. This advance imposes continuously growing requirements on the analyst. More and more new analytical procedures need to be developed for the selective or even specific determination of very minute amounts of various substances. In the field of metal analysis, the possibility of simple accomplishment of such tasks is offered primarily by organic complex forming reagents. These reagents form coordination compounds with metals, some particular characteristics of which permit their application in microanalysis.

The considerable development of coordination chemistry during the past two decades has made possible, *inter alia*, the recognition of the factors determining the analytical selectivity of complexation reactions. In the light of modern chemical knowledge, more and more analytical problems can be solved via complexation reactions.

For further progress in this field it is important that analytical chemists possess a coordination chemical view and apply this in their work. This is why the first part of this book deals with a coordination chemical treatment of analytical reactions based on complexation.

The second part of the book provides tried and time-honoured analytical procedures for the accomplishment of many metal analytical tasks. An effort has been made to present examples covering a wide range of applications in such a manner as to provide assistance in further problems not detailed here. The number of analytically applied organic reagents is so vast, and still rapidly increasing, that I could not aim at completeness in this respect. Instead, those methods are presented of which I myself have experience.

At the end of the book tabulated data facilitate experimental work in the field of metal analysis.

I sincerely hope that this book will provide useful aid both to those researchers engaged with further development in this field of analytical chemistry and to practical analysts busy with routine metal analyses.

The Hungarian edition of this book appeared early in 1969. The English edition has been supplemented by several newer procedures and further literary references.

I should like to express my sincere gratitude to Professor R. Belcher and Professor H. Freiser, Editors of this series of monographs, who initiated the edition of this book in English.

I thank Dr. Tivadar Mohácsy for translation of the text, Dr. D. Durham for stylistic revision of the translation and Miss Judit Temesi for her extensive technical assistance in preparation of this edition.

Finally I express my thanks to Mrs. E. K. Kállay for her thorough and conscientious editorial work.

KÁLMÁN BURGER

INTRODUCTION

The main trends in the development of analytical chemistry are determined by the demands of technical progress in the rapidly advancing old and the newly emerging branches of industry. This is why the determination of trace impurities in highly pure materials has become one of the central problems in analytical chemistry in this age of nuclear reactors and space vehicles, transistors and telecommunication satellites. The requirement is a double one: the *selective* determination of *very minute amounts of impurities* in the presence of several hundred or even thousand times as much of the main component. This duality is reflected in the two main directions of progress in analytical chemistry during the past twenty years: the development of instrumental methods and new analytical instruments which have increased the sensitivity of analytical procedures to an almost unbelievable degree; and the application, or the extension of the application, of selective and specific, highly sensitive organic reagents and reactions. These reagents[108, 258, 339, 473] (a) form the basis of selective analytical procedures, (b) make possible the combined separation and enrichment of trace constituents, and (c) may be employed as masking auxiliary complexing agents in analytical procedures.

The high sensitivity of organic reagents may be illustrated by the following examples: palladium(II) ions can be detected with *p*-dimethylaminobenzylidenerhodanine in a dilution of as great as 1 : 12,500,000. With the same reagent 0.02γ silver can be detected in a dilution of 1 : 2,500,000. Similar sensitivity is found in the determination of copper with rubeanic acid (0.006γ in a dilution of 1 : 2,500,000), or of lead with dithizone (0.04γ, 1 : 1,250,000), and in a number of similar estimations using organic reagents. If one notes that under optimum conditions the above reactions are also more or less specific, the paramount importance of organic reagents in the analysis of metals appears evident.

The great majority of the analytical eactions of organic reagents are based on complexation. A thorough examination and evaluation of these reactions has been made possible only by the recent developments in coordination chemistry. This is the reason that though sixty and eighty years have elapsed since the introduction of the first selective organic reagents: α-nitroso-β-naphthol[207] and dimethylglyoxime,[458] respectively, the physicochemical examination of the reactions utilized in analytical chemistry started as late as twenty years ago, while their evaluation has been made possible only by recent results.

The early investigations were mainly devoted to determining the stoichiometry and the quantitative course of reactions. In the next stage, the extent of the selectivity of reactions was determined by empirical "tests"; from many data the selectivity was held to be a property of the various functional groups and no attempts were made to find more fundamental explanations. The most outstanding representative of this era is Feigl.[158]

Contributions by coordination chemists,[226, 262, 283, 415, 453, 479, 483] opened up new vistas in research. Empirical tests were replaced by the exact physicochemical determination of stability constants and the thorough physical examination of the reaction products.

From a knowledge of the equilibrium constants of complex formation it is now possible in many cases to predict by simple calculation which metal ion or which ligand will interfere with a given complexation reaction,[378, 453] which ligand is suitable for masking a given metal ion in a given complexation reaction, and how this masking can be terminated later.[115, 379]

From a knowledge of the regularities of complex formation and the factors which determine the stability, solubility and the spectrum of a complex, it has become increasingly possible to design new organic reagents and to specify the scope and limitations of long-known reagents.[86, 96] Thus the study of the analytical application of organic reagents has emerged from the era of empirical data collection. More and more empirically elaborated analytical procedures become interpretable on a physicochemical basis and this greatly stimulates the development of new analytical procedures.

It should be mentioned, however, that though a knowledge of theoretical relationships considerably facilitates the design of the optimum conditions for analytical experiments and reduces the number of unsuccessful attempts, *it never substitutes for experimental work*. It holds true in analytical chemistry, too, that experiment is the criterion of theory. The results of research work carried out in this respect clearly indicate to analysts the importance of the contribution of coordination chemical investigations. On the other hand, it has frequently been proven that a thorough, detailed physicochemical study of reactions well known from an analytical viewpoint can result in the elucidation of complex chemical and bonding theory relationships.

Almost all organic reagents utilized in the analysis of metals are complex-forming agents. They give complexes with the metal ion to be determined or to be identified. The properties of the metal complex formed determine whether the organic ligand is suitable for application in gravimetry, volumetric analysis, spectrophotometry, polarography or in some other field of chemical analysis, and whether it should be used for the *enrichment* of the metal ion to be determined, the *masking* of interfering metal ions or the direct *determination* of the metal ion in question.

The reactions of organic reagents are *selective* if they react with a certain group of metals only, or if one or some of the complexes they give with various metal ions possess some particular physical or chemical properties. By means of selective reactions certain metals can be determined or detected in the presence of several other accompanying metal ions.

A selective reaction is termed *specific* provided it is characteristic of a single metal. However, although reactions considered as selective are usually common for many cations, by appropriate choice of the reaction conditions the selectivity can be enhanced in many cases, and under optimum conditions the reaction may even be made specific.

For many years Feigl and his co-workers[158] investigated how various organic compounds react with metal ions, and the nature of the relationship between the structure of an organic molecule and its reaction with metal ions. Their work resulted in a great number of new, more or less selective analytical reagents. After Feigl, many analysts[258, 329, 474] considered from an empirical viewpoint that the analytical selectivity of organic reagents is associated with the various functional groups they contain, but they usually did not look for more fundamental reasons.

The relationship between the basic strength of the ligand and complex stability and its analytical importance was first pointed out by Calvin[109a] in connection with the stabilities of cupric chelates of a number of enolic substances. The papers of Freiser[175] and Irving and Williams[228] presented at the First International Congress on Analytical Chemistry (Oxford, 1952) represent attempts to interpret analytical selectivity on a coordination chemical basis. Both papers emphasized the importance of the knowledge of the stability constants of complexes. They stated that by systematic variation of the factors affecting the stability of complexes, complex formation can be made selective or under optimum conditions even specific. Of these factors, both papers considered steric effects as the most promising from an analytical point of view. Irving stressed the role of orbital stabilization† and the effect of the basicity of the donor atom. In a continuation of Irving's work, Williams[479] called attention to the analytical significance of the thermodynamic relations of complex formation, and dealt with the role of the oxidation state and changes in the oxidation state of metals.

Szabó and Beck[453] provided a classical example for the application of the stability constants of complex formation and the acidic dissociation constants of the ligand in analytical chemistry. They demonstrated the possibility of calculating from such data those reaction conditions under which a given complex formation reaction is selective. During the last decade, Ringbom et al.[377-80] have dealt with the introduction of such calculation methods using the results and ideas of several complex chemistry research centres. It is due to their work that this approach is gaining more and more ground.

Erdey[148] interpreted the stability of complexes utilized for analytical purposes on the basis of the Lewis theory of acids and bases. He explained the Lewis acid strength of the central atom partly by electrostatic factors (ion potential, electronegativity) and partly on a quantum-chemical basis (bond types).

Bayer[35] pointed out the relationship between possible conjugation within the chelate ring and analytical selectivity.

The publications dealing with the problem illustrate the reasons for selectivity, with concrete examples (Irving:[228] the copper(I) complex of 2,9-dimethyl-1,10-phenanthroline; Bayer:[35] glyoxal-bis-(2-hydroxyanil) complexes, etc.).

Burger et al.[86] examined the reasons for the selectivity of dioximes, considered the most selective reagent group. Using equilibrium measurements,[35, 89, 94, 95, 148, 175, 377, 380] infrared,[98] ultraviolet and Mössbauer spectrometric analyses,[92] they established why dimethylglyoxime is specific for nickel(II) and palladium(II) ions.

These publications and those cited above clearly demonstrate that the question of the analytical selectivity of complex formation can be answered only by thorough coordination chemical examination of the individual systems. Since selectivity may be caused by different factors in the case of various complexes, the general

---

† By orbital stabilization is meant here the enhancement of stability due to electron pairing under the effect of strong-field ligands.

factors determining analytical selectivity may possibly become sufficiently clear from the study of numerous individual systems as to guide the investigation of other systems. Therefore, it seems necessary to the author to summarize these factors using the literature and his own experience.

It appears from the above that the analytical selectivity of these reactions is determined partly by the selectivity in complex formation and partly by the properties of the complexes formed. The former depends mainly upon the stability, the latter upon the electronic structure, symmetry, etc., of the respective complexes. The formation of mixed ligand complexes may have particularly high selectivity. Therefore, in Chaper 1 of this book we shall discuss the effects of stability and electronic structure of complexes, and of the formation of mixed ligand complexes upon the analytical selectivity.

CHAPTER 1

THE ANALYTICAL SELECTIVITY OF REACTIONS BASED ON COMPLEX FORMATION

I. EFFECT OF THE STABILITIES OF COMPLEXES UPON SELECTIVITY

A knowledge of the stabilities of metal complexes formed in aqueous solutions is very important for the analyst. He must know the stabilities of complexes formed by a given metal ion with various ligands and by a given ligand with various metal ions. A knowledge of the sequence of stability constants provides information as to which metal ions and anions (the latter acting as ligands) will interfere with determinations or detections carried out with a given ligand.

1. FACTORS DETERMINING THE STABILITIES OF COMPLEXES

The stabilities of metal complexes are determined by the physicochemical characteristics of the central metal ions and the ligands, mainly by their electronic structures and symmetry relations.

Effect of the metal ion

Depending on their charge, radius, electronic structure and polarizability, central metal ions form complexes of different stability with various ligands.

Sidgwick[430] classified metal ions into three groups according to the stabilities of their complexes with ligands containing oxygen and nitrogen donor atoms.

The metal ions belonging to group I form more stable complexes with ligands containing oxygen donor atoms than with ligands containing nitrogen donor atoms. These ions are the following: Mg(II), Ca(II), Sr(II), Ba(II), Co(II), Ga(II), In(III), Fe(III), Tl(III), Zr(IV), Th(IV), Si(IV), Ge(IV), Sn(IV), V(IV), V(V), Nb(V), Ta(V), Mo(V) and U(VI).

The metal ions of group II bind about equally strongly to oxygen and nitrogen: Be(II), Fe(II), Pd(II), Cr(III), Ru(III), Rh(III), Os(IV), Ir(IV) and Pt(IV).

Group III contains those metal ions which bind more strongly to ligands with nitrogen donor atoms than to ligands with oxygen donor atoms: Cu(I), Ag(I), Cu(II), Cd(II), Ni(II), Hg(II), V(III) and Co(III).

It is seen that group I consists mostly of metal ions with an inert gas configuration or containing few d-electrons, while group III involves metal ions with complete or almost complete d-shells. The greater readiness of the metal ions belonging to the latter group, to deform and the enhanced polarizability of the nitrogen atom, explain the greater affinities of these metal ions for nitrogen. The polarization effect of the ligand upon the central metal ion increases with a decrease in the charge and increase in the radius of the metal ion, and hence the greatest differences occur within the individual groups of the periodic system in groups IA (alkali metals) and IB (copper, silver, gold), whereas the complex-forming capabilities of the quadrivalent ions of groups IVA and IVB, for instance, are rather similar and they bind more strongly to oxygen than to nitrogen. The Fe(III) ion belongs to group I, but the other trivalent transition metal ions to the second and third groups. Due to its spherically symmetric d^5 electron configuration, the Fe(III) ion has no crystal-field stabilization energy[330] whereas in the case of the other metal ions mentioned this stabilization factor is significant. (See also in Section II, pp. 41–45.) Clearly, these latter ions bind more strongly to the nitrogen atom with its greater field-strength. Sidgwick examined the bond strengths between metal ions and oxygen and sulfur donor atoms, respectively. He found that while Be(II), Cu(II) and Au(III) ions bind more strongly to oxygen-containing ligands, Cu(I), Ag(I), Au(I), Hg(I) and Hg(II) ions prefer ligands with sulfur donor atoms.

The Sidgwick groups of metals can be further classified into sub-groups. Thus, for instance, within the group containing the Al(III), Ga(III), In(III) and Tl(III) ions, aluminium and gallium form rather ionic complexes (i.e. bound by electrostatic forces), while the forces acting in the complexes of indium and thallium are rather covalent. Thus these latter two ions form more stable complexes with ligands of greater polarizability, e.g. their complexes formed with iodide are more stable than the corresponding complexes formed with chloride ligands.

Ahrland, Chatt and Davies[3] classified metals into two groups according to their complex-forming behaviour. In this classification, group a contains those metal ions which give more stable complexes with light donor atoms (nitrogen, oxygen, fluorine) than with heavier donor atoms (phosphorus, sulfur, chlorine) in the same column of the periodic system. Conversely, group b contains those metal ions which give more stable complexes with the heavier donor atoms. In terms of this classification different ions of the same metal may belong to different classes, e.g. in their higher oxidation states the majority of metal ions belong to group a, whereas in their lower oxidation states many of the ions of the same metals belong to group b. For example, copper(I), silver(I), palladium(II), platinum(II), mercury(II), and transition metal in the oxidation state zero form complexes with ligands containing phosphorus, sulfur and chlorine donor atoms which are considerably more stable than those with light donor atoms. In the case of cobalt, nickel and copper complexes such a classification is much more difficult. At any rate, it can be stated that whenever the central metal ion is capable of back-coordination (metal → → ligand electron transfer)[84, 113, 485] the more stable complexes are formed with the heavy donor atoms. On the other hand, the role of back-coordination is the greater, in general, the lower the oxidation state and the ionization potential of the central metal ion. In such cases back-coordination makes possible the distribution of the excess negative charge on the low oxidation state central atom in the complex molecule, in accordance with the electroneutrality principle of Pauling.[335] This is why only ligands possessing π-acceptor ability stabilize metals in unusually low oxidation states.[330]

Thus, in the absence of steric and conjugation effects, the stability sequence with respect to the donor atoms of the complexes of metal ions belonging to group a is the following: $N \gg P > As > Sb$; $O \gg S$; $F \gg Cl > Br > I$. The reason for the difference in stability is the absence or negligible extent of back-coordination in the case of metals of group a; hence these metal ions form stable complexes with ligands whose donor atom is not a π-acceptor. On the other hand, metal ions capable of back-coordination with a metal $\rightarrow$ ligand electron transfer give more stable complexes with ligands containing π-acceptor donor atoms with unfilled p_π or d_π orbitals. It is worthy of mention that with mercury, thallium, tin and lead, for instance, the higher oxidation states are more capable of back-coordination because the lower oxidation state forms possess s-valency electrons which shield the d-electrons and hinder their participation in back-coordination.

Recently Pearson[336, 337, 454] introduced the concept of hard and soft Lewis acids and bases. In this classification, hard acids mean, in fact, the metal ions belonging to group a in Ahrland's classification,[3] while hard bases are the ligands with light, less polarizable donor atoms. Similarly, soft acids are the metal ions of group b (Ahrland), and soft bases the ligands with heavy, easily polarizable donor atoms. Pearson pointed out that hard (rigid) metal ions give stable complexes with hard (rigid) ligands and soft (polarizable) metal ions with soft (polarizable) ligands. This simple rule covers several, in some cases mutually contradictory, stabilizing factors, e.g. ionic and covalent σ-bonds, various types of π-bond, other interactions of the electronic system, solvation effects, etc. However, the above general rule is informative for the analyst and complex chemist even without a knowledge of the complicated factors of which it is composed.

The Irving–Williams stability sequence[227] provides valuable information on the stabilities of complexes. It gives the stability order of high spin complexes of the bivalent ions of the first-row transition metals as: $Mn < Fe < Co < Ni < < Cu < Zn$, independent of the participant ligands.

The greatest difference in stability occurs within this sequence between the manganese(II) and copper(II) complexes with ligands containing nitrogen donor atoms,[433] e.g. in the ethylenediamine complexes $\log \beta_2(Cu) - \log \beta_2(Mn) = 14.8$. The same difference is considerably smaller in the complexes of chelate-forming ligands, containing nitrogen and oxygen, e.g. it is 9.7 in the glycinate complexes. The difference is even smaller in complexes containing only oxygen donor atoms, for instance it is as low as 3.2 in the oxalate complexes. This great dependence upon donor atom results from the ligand-field stabilization: the field-strengths of ligands with nitrogen donor atoms are considerably greater than those with oxygen donor atoms.

Ligand-field stabilization is one of the most important factors which determine the stabilities of complexes formed by transition metal ions. It will be discussed briefly in Section II, subsection 1, together with the change in electronic structure of the central atom upon complex formation.

A number of other stability sequences have also been constructed empirically. Papers of Freiser et al.[130a, 161a, 287a, 423a] indicate the role of the sulfur donor atom in comparison with oxygen. A helpful correlation was also developed for the stabilities of nickel and zinc complexes. These types of empirical correlations can provide valuable information for the practical analyst.

In general, with less easily distorted central ions, e.g. the alkaline earth metals, the complex stability decreases with increasing radius of the cation. In the case of central ions with equal or nearly equal ionic radius and charge, the more distortable (polarizable) the central ion, the higher the stability of its complexes. Thus, for example, the size of the thallium(I) ion is between those of the potassium and rubidium ions, but since due to its closed d-shell it is more polarizable, it gives much more stable complexes with organic ligands.

Effect of the ligand

The overall stability constants (β) of related ligands with the same metal ion show the following linear relationship with the base strength (protonation constant) of the ligand (K_H):

$$\log \beta \sim a(-\log K_H) + b,$$

where a and b are constants. This relationship may be explained by the fact that the strength of the bond between a ligand and a metal ion is determined to a first approximation by the same factors as those which determine the strength of protonation of the ligand. Thus in general the basicity of a given ligand and the stability of its metal complex run parallel. According to Williams et al.[237] constant b is useful for the specification of the extent of the π-character of the bond. However, examination of the stabilities of copper complexes of substituted salicylaldehydes has revealed that constant b is dependent upon, inter alia, the position of substitution.[123] The electronic interactions may be considerably affected by substituents. Thus a substituent may influence, in addition to the basicity of the ligand, its π-acceptor capacity, resonance relations, etc.[90] Formation of a new ring within the ligand and opening of an existing one may also have significant effects. Thus, for example, when the ligand 2,2'-dipyridyl is replaced by 1,10-phenanthroline, the extra ring makes the ligand energetically more capable of accepting electrons donated by the central atom, i.e. of back-coordination. Though the structural change leaves the basicity, and therefore the tendency for establishment of a σ-bond, of the ligand almost intact, the change in π-acceptor capacity of the ligand makes the complex more stable.

In the case of complexes of chelating ligands, the chelation effect is one of the most important stabilizing factors.

Chelate formation is encountered whenever the ligand contains donor atoms so arranged that two or more donor atoms of the same ligand can bind to the same central atom. In this way the formation of chelate complexes is accompanied by the formation of one or more rings. Chelate complexes are considerably more stable than those containing monofunctional ligands. This is illustrated by the data in Table 1, which also show that the greater the number of rings formed in a chelate complex molecule, the higher is the stability of the complex. The M(en)$_2$, *Mtren* and *Mpenten* complexes contain two, four and five rings, respectively.

TABLE 1. LOGARITHMS OF THE STABILITY CONSTANTS OF SOME AMINE COMPLEXES AND OF POLY-AMINE CHELATES[59, 366, 414, 421]

Central atom	NH_3		Ethylenediamine (en)		Triamino-triethyl-amine (tren) $\log K$	Tetraamino-tetraethyl-ethylene-diamine (penten) $\log K$
	$\log \beta_4$	$\log \beta_6$	$\log \beta_2$	$\log \beta_3$		
Co(II)	5.31	5.29	10.9	13.11	12.8	15.8
Ni(II)	7.8	8.5	14.5	19.2	14.8	19.3
Cu(II)	12.59	—	20.2	20.3	18.8	22.4
Zn(II)	9.06	—	11.2	13.0	14.7	16.2
Cd(II)	6.92	4.92	10.3	12.4	12.3	16.8

By *chelate effect*[414] is meant the enhancement in stability due to ring formation; this is usually expressed as the difference between the logarithms of the stability constant of the chelate complex and the overall stability constant of the complex formed by the same metal ion with the corresponding monofunctional ligands. Table 2 shows that the chelate effect causes an increase in stability of several orders of magnitude.

TABLE 2. CHELATE EFFECT IN COMPLEXES OF SOME CHELATING AMINES ($\Delta \log K$)

Central atom	Ethylenediamine		Triethylene-tetramine	Tetraamino-tetraethyl-ethylene-diamine
	$\log \beta_2$	$\log \beta_3$		
Co(II)	5.6	7.8	7.5	10.5
Ni(II)	6.7	10.7	7.0	10.8
Cu(II)	7.6	—	6.2	9.8
Zn(II)	2.1	—	5.6	7.1
Cd(II)	3.4	7.5	4.4	11.9

The chelate effect depends on many factors; in the case of non-transition elements it is mainly an entropy effect, while with transition metals it is partly an enthalpy effect.

The extents of contribution to the chelate effect by entropy changes are nearly equal in analogous systems. This phenomenon is explained partly by the higher statistical probability of complex formation since the effective collision of fewer particles is required. Moreover, in the formation of chelate complexes a given ligand replaces two or more water molecules in the coordination sphere. Thus the number of independent particles increases during chelate formation; e.g. in the formation of EDTA complexes each molecule (Y) of EDTA replaces six water molecules:

$$M(H_2O)_6 + Y = MY + 6\,H_2O$$

This increases the disorder and therefore the entropy of the system. The entropy gain causes a considerable decrease in free energy, and this results in the increase in the stability of the complex.

The ligand-field stabilization effect too favours chelate formation. The field strengths of polydentate ligands are greater than those of monodentate ligands. The experiments of Spike and Parry[438] with metal complexes of ammonia and ethylenediamine illustrate this well. Ligand-field stabilization causes the increase in stability ensured by enthalpy changes in the chelate effect of transition metal chelates.

As a result of ring strain, the stabilities of chelate complexes depend greatly on the size of the chelate rings. Experience has shown that the stability is maximum, in general, with five- or at most six-membered rings. Exceptions are Ag(I) chelates in which the six-, seven- and eight-membered[416, 420] rings are more stable than the five-membered ones. The reason for this phenomenon is that with the increase in ring size the linear bonding most favoured by silver becomes more and more feasible.

The structure of the chelating ligand may affect the symmetry of the complex formed. For instance, the copper(II) ion forms planar or distorted octahedral complexes and thus it gives more stable complexes with triethylenetetramine, which is capable of forming complexes with such symmetry, than with triaminotriethylamine, which favours the formation of tetrahedral complexes, although in general complexes of the latter ligand are more stable (Table 3).

TABLE 3. LOGARITHMS OF THE STABILITY OF SOME TRIETHYLENETETRAMINE *(trien)* AND TRIAMINOTRIETHYLAMINE *(tren)* COMPLEXES[366, 413]

Central atom	trien	tren
Fe(II)	7.8	8.8
Co(II)	11.0	12.8
Ni(II)	14.0	14.8
Cu(II)	**20.4**	**18.8**
Zn(II)	12.1	14.7

Steric effects

It appears from the above that an appropriately chosen organic ligand (a selective reagent) can form more stable complexes with a certain group of metals than with others. However, *there is almost never such a great difference in the stabilities of complexes* formed by individual organic reagents and various metal ions of approximately identical electronic structure *as to make the reagent capable of the specific determination of a single metal.*

The selectivity of a given reagent can be enhanced by introduction of a suitable substituent which may be effective in various ways. It is the steric effect of the introduction of a new substituent which is the most important as regards analytical selectivity.

A requirement for the formation of stable complexes is an overlap between the unshared electronic orbitals of the central atom with the orbitals of those electrons of the ligand which are capable of being donated. Whenever some steric factor hinders this overlap, complex formation also is hindered and the stability of the product decreases. For instance, in the case of a monodentate ligand the donor atom of which has a large radius, steric hindrance may preclude the formation of complexes containing as many ligands as correspond to the maximum coordination number of the central atom. Steric hindrance is particularly significant with complexes of chelate-forming ligands and can largely determine the analytical selectivity of complexation. With these complexes, relatively minor changes in ligand structure may cause considerable changes in complex stability. A comparison of the complex-forming properties of 1,10-phenanthroline and 2,9-dimethyl-1,10-phenanthroline demonstrate this. The tris (1,10-phenanthroline) complex of Fe(II) is a low spin, extremely stable, vivid red compound. For steric reasons, three molecules of the dimethyl derivative of the ligand cannot enter the coordination sphere of the Fe(II) ion; the bis-complex, which is sterically possible, is a high spin, colourless compound of lower stability. The two methyl groups prevent the formation of the octahedral iron(II) complex, but they do not hinder the binding of two ligands to Cu(I) ions to give a tetrahedral complex. In these, the nucleophilic methyl groups increase the basicity of the ligand and thus the stability of the copper complex. Accordingly, appropriate substituents may prevent the arrangement of ligands around the central atom in the symmetry required for complexation. This explains, for instance, why the cuproine grouping

is specific for copper.[222, 223, 228] In copper(I) complexes the $4sp^3$ configuration makes necessary the tetrahedral arrangement of the two ligands; the substituents on the carbon atoms adjacent to the donor atoms do not interfere with this. However, they do hinder the simultaneous octahedral coordination of three ligands. Hence cuproine does not react with cations which form octahedral complexes of the type MA_3, or at most gives complexes with very low stabilities. Table 4 illustrates this with a comparison of the stabilities of some 1,10-phenanthroline *(ophen)* and 2,9-dimethyl-1,10-phenanthroline *(dphen)* complexes. It is seen that because of the steric hindrance exerted by the methyl groups, *dphen*, which has a considerable

TABLE 4. LOGARITHMS OF THE STABILITY CONSTANTS OF SOME 1,10-PHENATHROLINE *(ophen)* AND 2,9-DIMETHYL-1,10-PHENANTHROLINE *(dphen)* COMPLEXES[14, 437, 482]

Central atom	ophen			dphen
	$\log K_1$	$\log K_2$	$\log K_3$	$\log K_1$
H^+	4.95	—	—	6.15
Cd(II)	5.78	5.04	4.10	2.8
Ni(II)	8.8	8.3	7.7	2.6–2.8
Zn(II)	6.55	5.80	5.20	3.1

affinity for protons, forms complexes less stable by several orders of magnitude than does the less basic *ophen*.

The size of the central atom is also a factor of significance in steric effects. Thus, for instance, 2-methyl-8-hydroxyquinoline does not form a stable tris-complex with the aluminium(III) ion because of the low ionic radius of the metal, whereas in the case of the larger chromium(III), iron(III) and gallium(III) ions there is room in the coordination sphere of the central ion for the three ligands necessary to form a neutral complex. Because of the greater size of the phenyl group as compared with the methyl group, 2-phenyl-8-hydroxyquinoline does not form a stable complex with the chromium(III) ion either.

The steric hindrance exerted by substituents adjacent to the donor atom is illustrated by the comparison of the stabilities of complexes formed by Mg(II) with a series of substituted 8-hydroxyquinolines in Table 5. The steric hindrance due to the methyl group adjacent to the nitrogen donor atom is the most pronounced, but a methyl group close to the hydroxyl group also lowers the stability of the complex; the same substituent in position 5 increases the stability of the complex by increasing the basicity of the ligand.

TABLE 5. LOGARITHMS OF THE STABILITY CONSTANTS OF SOME MAGNESIUM(II) COMPLEXES OF MONOMETHYL-8-HYDROXYQUINOLINE DERIVATIVES.[225, 322]

Ligand	Mg(II)		
	$\log K_1$	$\log K_2$	pK_H
8-Hydroxy-2-methylquinoline	3.73	3.13	11.71
8-Hydroxy-5-methylquinoline	5.21	4.47	11.25
8-Hydroxy-6-methylquinoline	5.09	4.31	10.71
8-Hydroxy-7-methylquinoline	4.64	4.12	11.31
8-Hydroxyquinoline	5.04	4.29	10.80

Substituents can also exert a hindrance to complex formation by altering the symmetry of the ligand. Thus, for example, the introduction of two large groups in positions 3 and 3′ in 2,2′-dipyridyl prevents the planarity of the ligand molecule which in 2,2′-dipyridyl gives the possibility of resonance stabilization of the metal–ligand bond in the complex.[109]

In the case of ligands with the porphyrin skeleton, the size of the empty space in the centre of the ligand determines the choice of metal ions, with which complex formation can occur. Due to their large ionic radii, mercury(II) and lead(II) ions do not fit into this free space.[144] On the other hand, complexation with an ion of too low an ionic radius is hindered because the porphyrin skeleton is rigid and the donor atoms cannot approach the small central atom. This is the reason for the relatively high selectivity of such ligands.

2. THE RATIO OF STEPWISE STABILITY CONSTANTS, AND ANALYTICAL SELECTIVITY

From the investigations of Bjerrum[59] it is well known that the formation of complexes occurs stepwise:

$$M + A \rightleftarrows MA$$

$$MA + A \rightleftarrows MA_2$$

$$MA_{n-1} + A \rightleftarrows MA_n$$

With a knowledge of the equilibrium constants, i.e. the stepwise stability constants of the complexes, one can calculate the free ligand concentration (in the language of the analyst: the excess of reagent and the pH) at which the individual complexes form and what are the concentrations of species present in a solution of given total composition.

The organic reagents used in the analysis of metals are, in general, weak acids whose conjugated bases are capable of coordination as ligands. The concentration of the free ligand is determined by the acidic dissociation constant of the reagent, by the amount of reagent, and by the pH of the solution.

The ratio of stepwise stability constants generally obeys the following simple regularity: log K_n/K_{n+1} is positive and approximately constant. A different ratio of stepwise stability constants may be due to several reasons; it may indicate (1) a change in stereochemistry of the complex, (2) a transformation in its electronic structure (low spin → high spin), (3) the presence of π-bonds, (4) steric hindrance, (5) hydrogen bonds between ligands in the complex.

1. The $HgCl_2$ molecule is linear, while the $HgCl_4^{2-}$ complex is tetrahedral.[364] The originally sp hybrid is transformed to an sp^3 hybrid during the reaction $HgCl_2 +$ $+ Cl^- \rightarrow [HgCl_3^-]$ and this causes an anomalously large value of the ratio K_2/K_3. The reaction $FeCl_3 + Cl^- \rightarrow [FeCl_4^-]$ is accompanied by a change in the symmetry of the complex from octahedral to tetrahedral. The tetrahedral complex $[Cd(NH_3)_4^{2+}]$ is transformed during the uptake of a further ammonia molecule to the octahedral complex $[Cd(NH_3)_5H_2O^{2+}]$. These stereochemical changes are reflected in the stepwise stability constant ratios (Table 6).

2. Equilibrium measurement data[14] have revealed that $K_3 \gg K_1K_2$ for the 2,2'-dipyridyl and 1,10-phenanthroline complexes of iron(II). The great enhancement in stability which acccompanies the coordination of the third phenanthroline or dipyridyl ligand to the central iron atom is explained by the fact that while the MA_2 complexes are high spin (four unpaired electrons), the MA_3 complexes are low spin (diamagnetic).

TABLE 6. STEPWISE STABILITY CONSTANTS INDICATING STEREOCHEMICAL CHANGES OF COMPLEXES

Central atom	Ligand	log K_1	log K_2	log K_3	log K_4	log K_5	Refs.
Hg(II)	Cl^-	6.74	6.48	0.95	1.05	—	281
Fe(III)	Cl^-	1.48	0.65	1.0	—	—	367
Cd(II)	NH_3	2.65	2.10	1.44	0.93	−0.32	59

3. The significantly large first stepwise stability constant of the Ag(I) iodide complex is due to a donor π-bond arising from a metal → ligand electron transfer. The d-electron orbitals involved with the free π-orbitals of the first ligand in the donor π-bonds are much less available for the second ligand.[4] At the same time, the value of log K_2 is greater than log K_1 in the ammine complexes of silver, because the complex $[Ag(NH_3)_2^+]$ possesses the linear structure optimum for silver complexes, while in the complex $[Ag(NH_3)]$ the ammonia molecule replaces one of the water molecules of the tetrahedral or octahedral aquo complex.

4. In the case of the coordination of bulky ligands, steric effects may affect the ratio of stepwise stability constants. In Table 7 the log K_n/K_{n+1} values of nickel(II) complexes with ethylenediamine and some substituted derivatives are compared. It can be seen that the substituents hinder the coordination of the second and third ligands to the central metal ion and hence increase the values of log K_1/K_2 and log K_2/K_3.

TABLE 7. log K_n/K_{n+1} VALUES INDICATING STERIC HINDRANCE IN NICKEL(II) COMPLEXES OF ETHYLENEDIAMINE AND OF ITS N-SUBSTITUTED DERIVATIVES [33]

Ligand	log K_1/K_2	log K_2/K_3
$H_2N-CH_2-CH_2-NH_2$	1.12	1.45
$CH_3-HN-CH_2-CH_2-NH_2$	1.62	3.73
$CH_3-HN-CH_2-CH_2-NH-CH_3$	2.38	3.23
$C_2H_5-HN-CH_2-CH_2-NH-CH_3$	2.32	—

5. In complexes in which two or more ligands are linked by intramolecular hydrogen bonds in addition to their bonding to the central metal ion, $K_n \leq K_{n+1}$. Such hydrogen bonds occur, for example, in the chelates of dimethylglyoxime, salicylaldoxime and related ligands. The effect of hydrogen bonds is illustrated in Table 8 which shows the stability constants of some complexes of this type.

The stepwise stability constant ratios of complexes with ligands applied as analytical reagents are also of interest from an analytical point of view. In the following, as an example, the formation of complexes of a polyfunctional ligand (A) will be dealt with, two molecules of which fill the coordination sphere of the central atom. The same holds true *mutatis mutandis* for other systems as well.

Figure 1 shows the formation of a complex of the MA_2 type for four various ratios of the stepwise stability constants; the value of the overall stability constant ($\beta_2 = K_1K_2$) is 10^{17} in each case.

The ratio K_1/K_2 rarely corresponds to the statistical case shown in Fig. 1(a), particularly in the case of chelates. Figure 1(b) shows the more common case; this, however, is the least advantageous for the analyst; neither complex is formed quantitatively within a relatively broad range of log $[A^-]$. The complete formation of the complex MA_2 requires a free ligand concentration of over 10^{-5} M. Due to the weak acidities of the usual organic reagents (pK_a from 10^{-7} 10^{-11}), complex formation will be complete only at high pH values.

With a similar ratio of the stepwise stability constants and with higher values of the overall stability constant the situation does not change considerably;

Table 8. Stepwise Stability Constants of Some Complexes Indicating the Influence of Intramolecular Hydrogen Bonds

Central atom	Ligand	log K_1	log K_2	log β_2	Medium	Refs.
Mn(II)	dmg	8.6	8.6	17.2	50% dioxane	94
Co(II)	dmg	8.35	8.63	16.98	0.1 M NaClO$_4$	26
Ni(II)	dmg	$K_1 \ll K_2$		17.24	0.1 M NaClO$_4$	268
Cu(II)	dmg	11.0	12.5	23.5	50% dioxane	94
Zn(II)	dmg	8.1	9.3	17.4	50% dioxane	94
Pd(II)	dmg	$K_1 \ll K_2$		34.1	0.1 M NaClO$_4$	176
Mn(II)	CH$_3\cdot$Sa	6.14	6.14	12.28	75% dioxane	90
Mn(II)	Sa	5.8	6.1	11.9	75% dioxane	90
Mn(II)	Cl$\cdot$Sa	4.8	5.7	10.5	75% dioxane	90
Co(II)	CH$_3\cdot$Sa	6.82	7.52	14.34	75% dioxane	90
Co(II)	Sa	6.4	7.1	13.5	75% dioxane	90
Co(II)	Cl$\cdot$Sa	6.3	7.0	13.3	75% dioxane	90
Co(II)	NO$_2\cdot$Sa	6.3	6.6	12.9	75% dioxane	90
Ni(II)	CH$_3\cdot$Sa	7.2	7.5	14.7	75% dioxane	90
Ni(II)	Sa	6.9	7.4	14.3	75% dioxane	90
Ni(II)	Cl$\cdot$Sa	6.6	7.1	13.7	75% dioxane	90
Ni(II)	NO$_2\cdot$Sa	6.5	7.3	13.8	75% dioxane	90

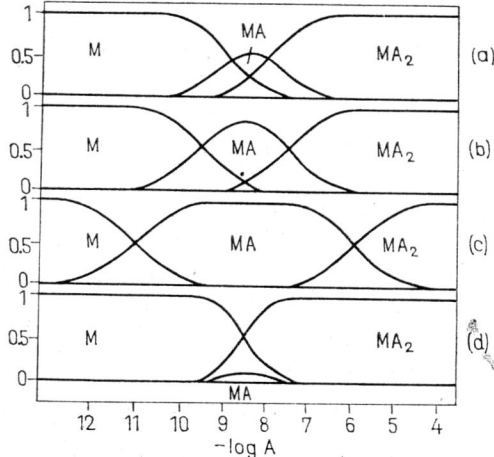

Fig. 1. Extent of formation of the successively formed species M, MA and MA$_2$ as a function of the logarithm of the free ligand concentration with various values of K_1/K_2. (a) The ratio of the successive stability constants corresponds to the statistical case: $K_1 = 4K_2$ ($K_1 = 10^{8.8}$; $K_2 = 10^{8.2}$). (b) The ratio of the successive stability constants is the usual, $K_1 > K_2$ ($K_1 = 10^{9.5}$, $K_2 = 10^{7.5}$). (c) Formation of the MA$_2$ complex is restricted because of steric reasons, configurational rearrangement or other causes, hence $K_1 \gg K_2$ ($K_1 = 10^{11}$, $K_2 = 10^6$). (d) Because of stabilizing effects (intramolecular hydrogen bonds, ligands of high field-strength, etc.) $K_2 > K_1$ ($K_1 = 10^{7.5}$, $K_2 = 10^{9.5}$)

the diagram analogous to that in Fig. 1(b) is merely shifted toward greater or smaller values of $-\log [A^-]$ depending on whether there is an increase or a decrease of stability, respectively.

The practical analyst knows well that the selectivity of reactions based on complexation in a neutral or alkaline medium is relatively low.

In the case shown in Fig. 1(c), $K_1 \gg K_2$ because of steric hindrance or configurational change. Therefore, an even higher concentration of the free ligand (higher pH) is necessary here for the complete formation of the complex MA_2; however, the formation of MA is complete within a relatively broad concentration range and this may be useful for analytical purposes.

The case shown in Fig. 1(d) represents the optimum situation from the point of view of the analyst. The conditions stabilizing the complex MA_2 (intramolecular hydrogen bonds, transformation of the electronic structure) ensure the quantitative formation of the complex even at low concentration of the free ligand (low pH). Therefore formation of MA_2 is much less dependent upon the excess of the reagent and the pH of the solution than in the previous cases. The analyst may choose the reaction conditions freely within broad limits, so as to make the reaction more selective or even specific under optimum conditions. It should be mentioned that the pH value which is optimum for the formation of a complex depends not only upon the stability of the complex. With a less basic ligand, the free ligand concentration necessary for complex formation may be reached at a lower pH than in the case of more basic ligands. However, complex stability usually decreases in parallel with the basicity of the ligand. Hence the formation of complexes with less basic ligands requires higher concentrations of the free ligand.[268] Consequently, the analyst can more advantageously apply complexes of less basic ligands the stabilities of which are enhanced by other factors (hydrogen bond, back-coordination, etc.), thus compensating the decrease in stability due to the weak basic character of the ligand.

From the above considerations it may be stated that whenever the analytical procedure is based on the complex MA_2, the case $K_1/K_2 < 1$ is the most advantageous for the analyst.

3. THE SIGNIFICANCE OF STABILITY CONSTANTS AND EQUILIBRIUM CALCULATIONS IN DETERMINING THE ANALYTICAL SELECTIVITY

From a knowledge of the equilibrium constants of complexation reactions, from the stability constants of complexes, one can calculate the extent to which a reaction proceeds with a given excess of the ligand, and the ratio of free and complexed metal. Hence these calculations provide information on the accuracy of analytical reactions based on complexation.

A knowledge of the stability constants of metal complexes enables us to predict which metal or metals will react with the ligand available in a solution containing several different metal ions and the one ligand[453] and also the concentrations of the various metal complexes in the solution. Similar information can be gained about the complexes present in a solution of one metal ion and several ligands.[176]

Thus, these calculations provide a measure of the selectivity of the given complexation reactions.

Whenever the ligands examined are relatively strong bases (anions of weak acids), acidic dissociation equilibria interfere with these calculations. A consideration of such interference requires the knowledge of the acidic dissociation constants. If ligands of several types are simultaneously present in the solution, mixed ligand complexes also may form. At higher metal concentrations the probability of processes leading to the formation of polynuclear complexes increases. With a knowledge of the respective equilibrium constants, all these equilibria can be taken into account. Many equilibrium constants determined by coordination chemists are now available to the analyst in tabular form.[433] Their practical application, however, is rather difficult, since in analytical practice those optimum conditions almost never occur when only one or at most two or three equilibria determine which complexes will be formed and in which concentration. Thus, for instance, if one calculates from the solubility product of mercury sulfide the amount of Hg(II) not precipitated in the gravimetric determination of Hg(II) ions as mercury sulfide, and hence the error of the determination, with a 0.1 N excess of the sulfide ion (the precipitating agent), one obtains a result of 10^{-51} mole/litre. Nevertheless, experimental data show that the solubility of mercury sulfide in 0.1 M sodium sulfide solution is 8×10^{-3} mole/litre. The striking difference between calculated and measured values becomes understandable if one considers the formation of the species $Hg(SH)_2$, $[HgS_2^{2-}]$, etc., which are soluble in water. The situation would be even more complicated in the titration with ethylenediaminetetraacetic acid (EDTA) of a metal ion capable of the formation of ammine complexes if one wished to calculate the concentration of free metal ions simply from the stability constant of the EDTA complex to be formed. The metal ion may exist in the ammoniacal medium as various ammine or even hydroxo complexes which form stepwise. In addition, depending upon the pH of the solution, the EDTA ligand is protonated to various extents.

As is clear from the above, the application of thermodynamic equilibrium constants determined by complex chemists under optimum conditions is made difficult by the emergence of secondary reactions in the usual analytical systems and these lead to products disregarded in the main reaction. The analyst is not interested in the concentration of each species present, but only in the extent to which the main reaction proceeds (the quantitativeness of the analytical reaction). For example, the error of analytical sulfide precipitation methods is given by the amount of metal which remains in solution, irrespective of the form in which it exists. When a metal ion is titrated with EDTA in ammoniacal medium, we must know that fraction of the metal cation which reacted with the standard EDTA solution, but it is quite immaterial whether the unreacted fraction of the metal ion exists as free ion, or ammine or hydroxo complex, and whether the unreacted EDTA is protonated or not and to what extent.

The conditional equilibrium constants and their application

The situation becomes simplified for the analyst by the application of *conditional equilibrium constants*. Schwarzenbach,[415] Yatsimirskii,[484] Wehler,[470] Ringbom[377-80] and others suggested the introduction of conditional equilibrium constants to analytical calculations. For instance, in the cases dealt with above, the equilibrium constants and conditional equilibrium constants are as follows:

1. The solubility product K_{So} of mercury sulfide is:

$$K_{So} = [Hg^{2+}][S^{2-}]. \qquad (1)$$

The conditional solubility product is

$$[K'_{So}] = [Hg'][S'], \qquad (2)$$

where [Hg'] is the concentration of mercury not precipitated by the sulfide anion, irrespective of the form in which it exists in the solution, and [S'] is the concentration of the sulfide ion which is not bonded to mercury; that is

$$[Hg'] = [Hg^{2+}] + [Hg(SH)_2] + [Hg(S)_2^{2-}] + \ldots \qquad (3)$$

and

$$[S'] = [S^{2-}] + [HS^-] + [H_2S]. \qquad (4)$$

2. The stability constant of the complex formed during the titration of Zn(II) ions with EDTA is

$$K = \frac{[ZnY]}{[Zn^{2+}][Y]}, \qquad (5)$$

where Y is the EDTA ion with four negative charges.

The conditional stability constant is

$$K' = \frac{[ZnY]}{[Zn'][Y']}, \qquad (6)$$

where

$$[Zn'] = [Zn^{2+}] + [Zn(NH_3)^{2+}] + [Zn(NH_3)_2^{2+}] +$$
$$+ [Zn(NH_3)_3^{2+}] + [Zn(NH_3)_4^{2+}] + [Zn(OH)^+] + [Zn(OH)_2] + \ldots \qquad (7)$$

and

$$[Y'] = [Y^{4-}] + [HY^{3-}] + [H_2Y^{2-}] + [H_3Y^-] + [H_4Y]. \qquad (8)$$

Therefore the conditional stability constant relates those quantities in which the analyst is interested: the concentration of product ZnY, the concentration of the unreacted metal [M'] and the concentration of the unreacted EDTA [Y'].

Calculations can be performed with the above and analogous conditional equilibrium constants just as with thermodynamic equilibrium constants. Thus, one can calculate the solubility of the given precipitate, the concentration of the metal ion at the equivalence point of the titration based on complexation, etc., provided that the values [S'] and [Y'] are known. In this way, therefore, relatively quite complicated systems can be treated with ease.

Calculation of conditional equilibrium constants

It follows from the above that the determination of conditional equilibrium constants is of importance to the analyst. Of course, the value of conditional equilibrium constants depends on the pH of the solution, the concentration of ligands or metal ions which participate not in the main reaction but in side reactions, etc. The application of conditional stability constants would be of little benefit, if its calculation or determination were complicated or troublesome. Introduction of the coefficient α, employed first by Schwarzenbach, to specify the extent to which secondary processes proceed, however, makes the situation quite simple. For the preceding examples

$$\alpha_{Hg} = \frac{[Hg']}{[Hg^{2+}]} \tag{9}$$

and

$$\alpha_S = \frac{S'}{[S^{2-}]} \tag{10}$$

and

$$\alpha_{Zn} = \frac{Zn'}{[Zn^{2+}]} \tag{11}$$

and

$$\alpha_Y = \frac{[Y']}{[Y]} . \tag{12}$$

If the reactants participate only in the main reaction, $\alpha = 1$; whereas if at least one of them also takes part in some side reaction, $\alpha > 1$.

Accordingly, with a knowledge of the α values, the conditional equilibrium constants are calculated from equations (1), (2) and (9), (10) or (5), (6) and (11), (12) using the following, simple relationships:

$$K'_{So} = K_{So} \alpha_{Hg} \alpha_S \tag{13}$$

and

$$K' = \frac{K}{\alpha_{Zn} \alpha_Y} . \tag{14}$$

Analogous equations can be deduced for any similar equilibria.

The α values are functions of the equilibrium constants of the side-reactions. It can easily be shown, for instance, that in the case of the titration of zinc dealt with above

$$\alpha_{Zn} = \frac{[Zn^{2+}] + [Zn(NH_3)^{2+}] + [Zn(NH_3)_2^{2+}] + [Zn(NH_3)_3^{2+}] + [Zn(NH_3)_4^{2+}]}{[Zn^{2+}]} . \tag{15}$$

By expressing the concentration of complexes with stepwise and overall stability constants, respectively, and by subsequent reduction, we have

$$\alpha_{Zn} = 1 + K_1[NH_3] + \beta_2[NH_3]^2 + \beta_3[NH_3]^3 + \beta_4[NH_3]^4, \tag{16}$$

where

$$K_1 = \frac{[Zn(NH_3)^{2+}]}{[Zn^{2+}][NH_3]}; \quad \beta_2 = \frac{[Zn(NH_3)_2^{2+}]}{[Zn^{2+}][NH_3]^2};$$

$$\beta_3 = \frac{[Zn(NH_3)_3^{2+}]}{[Zn^2][NH^+_3]^3}; \quad \beta_4 = \frac{[Zn(NH_3)_4^{2+}]}{[Zn^{2+}][NH_3]^4} \tag{17}$$

and

$$\alpha_Y = \frac{[Y^{4-}] + [HY^{3-}] + [H_2Y^{2-}] + [H_3Y^-] + [H_4Y]}{[Y^{4-}]}. \tag{18}$$

Similarly, if the concentrations of protonated species are expressed with protonation constants, subsequent reduction gives the following relation:

$$\alpha_Y = 1 + K_1[H^+] + \beta_2[H^+]^2 + \beta_3[H^+]^3 + \beta_4[H^+]^4, \tag{19}$$

where

$$K_1' = \frac{[HY^{3-}]}{[H^+][Y^{4-}]}; \quad \beta_2 = \frac{[H_2Y^{2-}]}{[H^+]^2[Y^{4-}]}; \quad \beta_3 = \frac{[H_3Y^-]}{[H^+]^3[Y^{4-}]};$$

$$\beta_4 = \frac{[H_4Y]}{[H^+]^4[Y^{2-}]}. \tag{20}$$

Accordingly, from the above equations one can simply calculate the α values for a given concentration of the components involved in side-processes. Thus, one can plot the α values of metals capable of forming ammine complexes vs. the concentration of ammonia; the α values of weak acids vs. pH of the medium; etc. (Figs 2 and 3, respectively). Using the α values read from these and analogous graphs, it is easy to determine the extent to which a given reaction proceeds in solutions containing ammonia (or other complexing agent) in a given concentration and at a given pH.

The above considerations and in particular equations (16) and (19) show the manner in which the analyst can apply thermodynamic equilibrium constants most profitably. From the determined α values he can calculate the value or values of conditional equilibrium constants for the given conditions and these, similar to the α values, can be plotted against the ammonia concentration or pH (Fig. 4). The conditional equilibrium constant for a given concentration of ammonia (or other complexing agent) and a given pH can be applied for the calculation of

TABLE 9. LOG α VALUES OF SOME MORE IMPORTANT LIGANDS USED AS MASKING AGENTS AS A FUNCTION OF pH

pH	EDTA	NH$_3$	NTA	CH$_3$COO$^-$	Citrate	CN$^-$	PO$_4^{3-}$	P$_2$O$_7^{4-}$	Tartrate
0	21.4	9.4	14.4	4.65	13.5	9.2	20.7	18.1	7.0
1	17.4	8.4	11.4	3.65	10.5	8.2	17.7	14.4	5.0
2	13.7	7.4	8.7	2.65	7.5	7.2	15.0	11.3	3.05
3	10.8	6.4	7.0	1.66	4.8	6.2	12.65	8.7	1.4
4	8.6	5.4	5.8	0.74	2.7	5.2	10.6	6.6	0.4
5	6.6	4.4	4.8	0.16	1.2	4.2	8.6	4.6	0.05
6	4.8	3.4	3.8	0.02	0.25	3.2	6.65	2.9	—
7	3.4	2.4	2.8	—	0.05	2.2	5.0	1.6	—
8	2.3	1.4	1.8	—	—	1.2	3.7	0.6	—
9	1.4	0.5	0.9	—	—	0.4	2.7	0.1	—
10	0.5	0.1	0.2	—	—	0.1	1.7	—	—
11	0.1	—	—	—	—	—	0.8	—	—
12	—	—	—	—	—	—	0.2	—	—
13									
14									

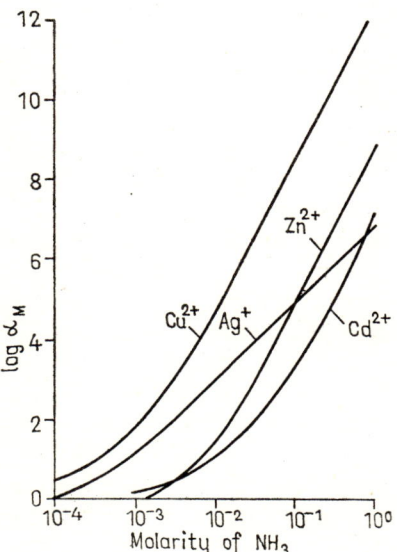

FIG. 2. Dependence of the log α values of some ammine-forming metal ions on the concentration of ammonia

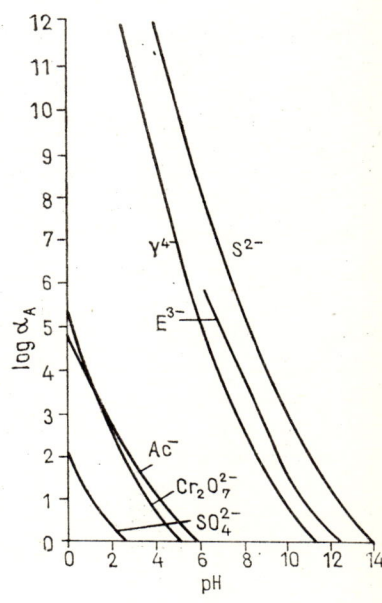

FIG. 3. Dependence of the log α values of weak acids on the pH (Ac = acetate, Y^{4-} = EDTA, E^{3-} = eriochrome black)

TABLE 10. Log α Values for $3d^5$–$3d^{10}$ Transition Metal Ions in the Presence of Some Important Masking Ligands, as a Function of pH. (Ligand concentration: 0.1 M; for ammonia: 1 M)

Central atom	Ligand	pH 1	2	3	4	5	6	7	8	9	10	11	12	13	14
Mn(II)	Citrate	—	—	0.1	0.7	1.5	2.1	2.4	2.4	2.4	2.4	2.4	2.4	2.7	3.4
	EDTA	—	0.5	2.4	4.3	6.3	8.0	9.4	10.4	11.4	12.2	12.4	12.4	12.4	12.4
	NTA	—	—	0.1	0.5	1.3	2.3	3.3	4.3	5.2	5.9	6.0	6.0	6.0	6.0
	OH$^-$	—	—	—	—	—	—	—	—	—	0.1	0.5	1.4	2.4	3.4
Fe(II)	Citrate	—	—	—	0.5	2.6	4.2	5.5	6.5	7.5	8.5	9.5	10.5	11.5	12.5
	EDTA	—	0.6	2.6	4.6	6.6	8.3	9.7	10.7	11.7	12.5	12.7	12.8	12.8	12.8
	NTA	—	—	0.7	1.7	2.7	3.7	4.7	5.7	6.6	7.4	7.9	8.8	9.8	10.8
	OH$^-$	—	—	—	—	—	—	—	—	0.1	0.6	1.5	2.5	3.5	4.5
Fe(III)	Citrate	0.3	3.0	6.4	9.5	12.0	13.7	15.0	16.0	17.0	18.0	19.0	20.0	21.0	22.0
	CH$_3$COO$^-$	—	0.2	1.3	3.5	5.2	6.0	7.7	9.7	11.7	13.7	15.7	17.7	19.7	21.7
	EDTA	7.0	10.3	13.0	15.2	17.2	18.9	20.4	22.0	24.0	26.4	28.5	30.6	32.6	34.6
	NTA	3.2	5.8	8.2	10.3	12.3	14.3	16.3	18.3	20.1	21.5	21.8	22.4	23.3	24.3
	OH$^-$	—	—	0.4	1.8	3.7	5.7	7.7	9.7	11.7	13.7	15.7	17.7	19.7	21.7
	P$_2$O$_7^{4-}$	—	6.0	10.6	13.8	16.0	18.0	19.4	20.0	—	—	—	—	—	—
Co(II)	Citrate	—	—	0.5	1.5	2.5	3.1	3.4	3.8	4.5	5.5	6.5	7.5	8.5	10.3
	EDTA	0.2	2.6	4.7	6.6	8.6	10.3	11.7	12.7	13.7	14.5	14.7	14.8	14.8	14.8
	NH$_3$	—	—	—	—	—	—	0.2	1.2	3.7	5.3	5.7	5.8	7.2	10.2
	NTA	—	0.8	2.4	3.5	4.5	5.6	7.1	9.0	10.8	12.2	12.4	12.4	12.4	12.4
	Tartrate	—	—	0.2	0.8	1.1	1.1	1.1	1.1	1.1	1.1	2.2	4.2	7.2	10.2

ANALYTICAL SELECTIVITY OF REACTIONS

Ni(II)	Citrate	—	0.2	0.5	1.7	2.9	3.6	4.4	5.3	6.3	7.3	8.3	9.3	10.3	11.3
	CH$_3$COO$^-$	—	—	—	—	0.1	0.2	0.2	0.2	0.3	0.7	1.6	—	—	—
	CN$^-$	—	5.1	2.5	6.5	10.5	14.5	18.5	22.5	25.7	26.9	27.3	27.3	27.3	27.3
	EDTA	2.4	—	7.1	9.0	10.9	12.6	14.0	15.0	16.0	16.8	17.0	17.1	17.1	17.1
	NH$_3$	—	—	—	—	—	0.1	0.6	2.8	6.3	8.3	8.8	8.8	8.8	8.8
	NTA	—	1.4	3.1	4.2	5.2	6.5	8.2	10.2	12.0	13.4	13.6	13.6	13.6	13.6
	OH$^-$	—	—	—	—	—	—	—	—	0.1	0.7	1.6	—	—	—
Cu(II)	Citrate	—	0.2	0.8	2.9	5.1	6.7	8.0	9.0	10.0	11.0	12.0	13.0	14.0	15.0
	CH$_3$COO$^-$	1.9	—	—	0.3	0.9	1.1	1.1	1.1	1.3	1.8	2.7	3.7	4.7	5.7
	EDTA	—	4.6	6.8	8.7	10.7	12.5	13.9	15.0	16.0	16.8	17.3	18.0	18.9	19.9
	NH$_3$	0.3	—	—	—	—	0.2	1.2	3.6	6.7	8.2	8.6	8.6	8.6	8.6
	NTA	—	2.6	4.3	5.5	6.5	7.5	8.8	10.5	12.3	13.9	14.1	14.3	15.1	16.1
	OH$^-$	—	—	—	—	—	—	—	0.2	0.8	1.7	2.7	3.7	4.7	5.7
	P$_2$O$_7^{4-}$	—	—	—	0.1	1.1	2.8	4.3	5.9	6.8	7.0	7.0	7.0	7.0	7.0
	Tartrate	—	0.1	1.0	2.5	3.1	3.2	3.2	3.2	3.2	3.2	3.3	3.8	4.7	5.7
Zn(II)	Citrate	—	—	0.3	1.4	2.6	3.2	3.5	3.8	4.4	5.4	6.4	8.5	11.8	15.5
	CH$_3$COO$^-$	—	—	—	0.1	0.5	0.6	0.6	0.6	0.7	2.4	5.4	8.5	11.8	15.5
	CN$^-$	—	—	—	—	—	0.1	3.5	7.5	10.7	12.3	12.7	12.7	12.8	15.5
	EDTA	0.3	2.8	4.9	6.8	8.8	10.5	11.9	12.9	13.9	14.7	14.9	15.0	15.0	15.6
	NH$_3$	—	—	—	—	—	—	—	0.4	3.2	4.7	5.6	8.5	11.8	15.5
	NTA	—	0.7	2.3	3.4	4.4	4.4	6.4	7.4	8.3	9.0	9.1	9.2	11.8	15.5
	OH$^-$	—	—	—	—	—	—	—	—	0.2	2.4	5.4	8.5	11.8	15.5
	Tartrate	—	0.2	1.4	2.4	2.8	2.8	2.8	2.8	2.8	2.8	5.4	8.5	11.8	15.5

the compositions (concentration of the complex, etc.) of such solutions, in which it is indeed constant (as has been calculated), similar to the thermodynamic constants in the usual complex equilibrium calculations. If more than one secondary process is involved, one has to calculate with the sum of the individual α values. Tables 9 and 10 give α values referring to a few metals and masking agents which are of importance in analytical practice. Similar data for other systems can be calculated from a knowledge of the respective thermodynamic equilibrium constants, using equations (16) and (19) or analogous equations.

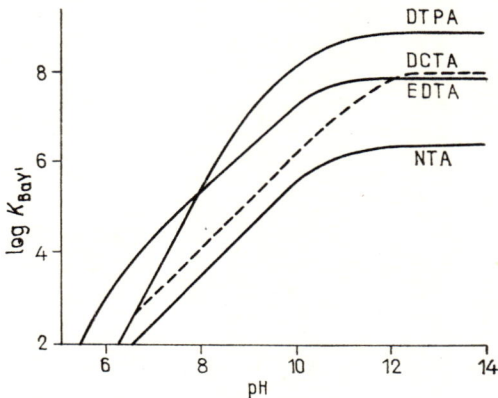

FIG. 4. Dependence of the conditional stability constants of some barium complexes on the pH (DTPA = diethylene triaminepentaacetate; DCTA = 1,2-diaminocyclohexanetetraacetate; EDTA = ethylenediamine tetraacetate; NTA = nitrilotriacetic acid)

Consideration of the formation of protonated and hydroxo mixed complexes

With the conditional stability constant defined by equation (6), we have taken into account only those side-reactions which affect the concentrations of either the ligand or the metal ion. With the calculations usual in analytical practice, such constants are applied in general with satisfactory results. However, with chelate complexes, one must take into consideration the formation of metal complexes containing the protonated ligand; these also participate in the analytical reaction. Similarly, hydroxo mixed complexes occur too in almost all complex systems. Thus, for instance, in the course of titrations with EDTA, complexes M(HY) and M(OH)Y may be formed in acidic and alkaline media, respectively. Since the metal:ligand ratio is 1:1 in these complexes as in the complex MY, to the analyst the formation of protonated or hydroxo mixed complexes is without significance from a stoichiometric point of view. Nevertheless, as the presence of these protonated or hydroxo mixed complexes necessarily shifts the equilibria, we must take their presence into consideration in the calculation of conditional equilibrium constants whenever the constants are used in calculations connected with measurements made when these species may be present. Therefore, in such cases the α values referring to protonated or mixed complexes must also be introduced:

$$\alpha_{MY} = \frac{[MY']}{[MY]}. \tag{21}$$

In acidic medium

$$[MY'] = [MY] + [MHY] = [MY](1 + K[H^+]), \qquad (22)$$

where

$$K = \frac{[MHY]}{[MY][H^+]}. \qquad (23)$$

In alkaline medium

$$[MY'] = [MY] + [M(OH)Y] = [MY](1 + K[OH]), \qquad (24)$$

where

$$K = \frac{[M(OH)Y]}{[MY][OH]}. \qquad (25)$$

Therefore, after consideration of the secondary reaction leading to the formation of the other product, the apparent equilibrium constant has the form:

$$K' = \frac{[MY']}{[M'][Y']} = \frac{\alpha_{MY} K}{\alpha_M \alpha_Y}. \qquad (26)$$

Masking and demasking

Masking is the transformation (without physical isolation) of components present in the sample in addition to the main component to be determined such that the transformed components do not give their characteristic reactions. For instance, cyanide ions react with zinc ions to give a highly stable complex which does not react with EDTA or the indicator eriochrome black T. The lead(II) content of the solution can be determined accurately under the same conditions, as this latter cation does not form a stable complex with cyanide.

Demasking is the recovery from the masked ion of its original form. For example, the zinc ions are released from their cyanide complex by addition of formaldehyde which reacts with the cyanide to give formaldehyde cyanohydrin and its decomposition products. The released zinc is then capable of its normal reactions with the above complexing agents.

In analytical practice, the reactions based on complexation are generally accomplished in the presence of other ligands (buffer, metal indicator, masking agent, etc.) in addition to the complexing agent. As has been shown above, the side reactions of these additives decrease the conditional equilibrium constant of the complexation process which forms the basis of the analytical procedure.

Therefore, masking can be characterized quantitatively by the decrease in conditional equilibrium constant. The α values serving for the calculation of the conditional equilibrium constants are quantitative measures of masking. Figure 5 demonstrates the masking effect of hydrogen ion concentration upon zinc

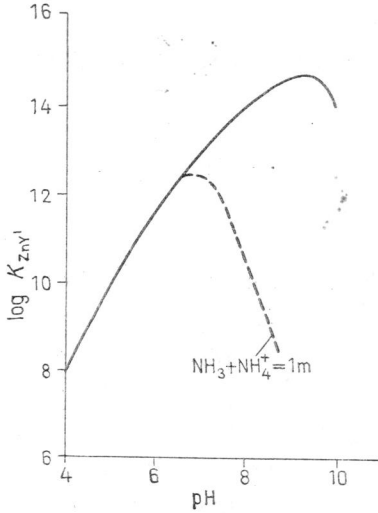

Fig. 5. Dependence of the conditional stability constant of the zinc(II) complex of EDTA on the pH (in the presence and absence of 1 M ammonia buffer)

ions in the titration with EDTA, and also the masking effect of ammonia concentration in the same system. Figure 5 illustrates the conditional stability constant of the EDTA complex of zinc as a function of the pH in the presence and absence of a 1 M ammonia buffer.

Thus, the conditional equilibrium constant provides information about the extent of masking too. According to Ringbom,[379] a reaction may be considered completely masked if the value of the conditional equilibrium constant is less than 100. On the other hand, the quantitative occurrence of the reaction may be expected only if the logarithm of the conditional equilibrium constant is greater than 7.

In order to specify the extent to which masking reactions and the analytically used complexation reactions proceed (the latter in the presence of masking complexing agent), Cheng[115] suggested the introduction of selectivity and masking factors:

$$\text{Selectivity factor } (S.F.) = \frac{(-\log M)^2}{-\log M_A}$$

$$\text{Masking factor } (M.F.) = \frac{(-\log M_A)^2}{-\log M}$$

where M is the concentration of free metal ions originating from the dissociation of the complex formed in the main reaction; and M_A is the concentration of free metal ions from the dissociation of the complex formed with the masking reagent.

If in addition to the metal ions and the ligand necessary for the complexation reaction the solution also contains a masking reagent, the occurrence of the main reaction is the more favoured, the greater the selectivity factor, while the effect of the masking reagent is the more expressed, the less this value. To illustrate the situation, Cheng[115] presented the values of selectivity and masking factors for a number of complexation reactions (Tables 11 and 12). The values of these factors evidently depend upon the concentration and concentration relations of the reagents, since any equilibrium can be shifted, in principle, in either direction by the addition of one of the reactants or one of the products in an appropriately high excess. On the other hand, the solubility relations of the reactants limit this possibility. For example, 10^{-3} mole of silver chloride can be dissolved in 1 liter of water by the addition of 0.025 mole of ammonia, whereas from calculations with the conditional equilibrium constants the dissolution of silver iodide would require 27 moles of ammonia. A 27 M ammonia solution cannot be prepared, but a 0.025 M solution presents no difficulty. The precipitation of silver chloride can easily be masked with ammonia, that of silver iodide not. The precipitation of silver chloride can also be masked with EDTA, provided that the concentration relations are appropriately selected. The selectivity factor of silver chloride is 7 in this particular case. Similarly, the copper(II) ion can be masked with EDTA against dimethylglyoxime; the selectivity factor is 6.5, and the masking factor 11.3. The formation of copper dimethylglyoxime cannot be masked with N,N-bis-(2-hydroxyethyl)-glycine since in the presence of the latter the selectivity

TABLE 11. STABILITY CONSTANTS K AND SOLUBILITIES S OF SILVER COMPLEXES AND THEIR SELECTIVITY FACTORS *(S.F.)* AND MASKING FACTORS *(M.F.)* IN THE PRESENCE OF EDTA

Ligand	Equilibrium	K	S	M.F.	S.F.
Ammonia	$Ag^+ + 2\,NH_3 \rightleftharpoons Ag(NH_3)_2^+$	1.5×10^7			
1,2,3,-Benzo- triazole	$Ag^+ + C_6H_4N_3^-$ $\rightleftharpoons AgC_6H_4N_3$		10^{-14}	1.85	13.6
Bromide	$Ag^+ + Br^- \rightleftharpoons AgBr$		5×10^{-13}	2.12	10.3
Chloride	$Ag^+ + Cl^- \rightleftharpoons AgCl$		1.0×10^{-10}	2.59	7.0
Chromate	$2\,Ag^+ + CrO_4^{2-} \rightleftharpoons Ag_2CrO_4$		1.9×10^{-12}	3.6	3.6
Cyanide	$Ag^+ + 2\,CN^- \rightleftharpoons Ag(CN)_2^-$	5.6×10^{18}		2.02	11.4
EDTA	$Ag^+ + EDTA \rightleftharpoons AgEDTA^{2-}$	2×10^7			
Hydroxyl	$2\,Ag^+ + 2\,OH^- \rightleftharpoons Ag_2O + H_2O$		2×10^8	3.41	4.0
Iodate	$Ag^+ + IO_3^- \rightleftharpoons AgIO_3$		2×10^{-8}	3.41	4.0
Iodide	$Ag^+ + I^- \rightleftharpoons AgI$		8.5×10^{-17}	1.62	17.8
Oxalate	$2\,Ag^+ + C_2O_4^{2-} \rightleftharpoons Ag_2C_2O_4$		5×10^{-12}	3.73	3.4
p-Dimethyl- amino benzylidene- rhodanine	$Ag^+ + C_{12}H_{11}OS_2N_2$ $\rightleftharpoons AgC_{12}H_{11}OS_2N_2$		8.1×10^{-19}	1.36	25.1
Sulfide	$2\,Ag^+ + S^{2-} \rightleftharpoons Ag_2S$		1×10^{-50}	0.79	73.8
Thiocyanate	$Ag^+ + SCN^- \rightleftharpoons AgSCN$		1×10^{-12}	2.16	10.0
Thiosulfate	$Ag^+ + 2\,S_2O_3^{2-} \rightleftharpoons Ag(S_2O_3)_2^{3-}$	1.7×10^{13}		2.82	5.9

TABLE 12. SELECTIVITY (S.F.) AND MASKING FACTORS (M.F.) OF
DIMETHYLGLYOXIME COMPLEXES IN THE PRESENCE OF EDTA AND BHEG

Metal	Masking agent	M.F.	S.F.
Co^{2+}	$EDTA^{4-}$	10.4	4.9
Co^{2+}	$BHEG^-$	3.1	9.0
Cu^{2+}	$EDTA^{4-}$	11.3	6.5
Cu^{2+}	$BHEG^-$	5.7	8.6
Ni^{2+}	$EDTA^{4-}$	12.0	5.6
Ni^{2+}	$BHEG^-$	4.1	9.6
Zn^{2+}	$EDTA^{4-}$	14.3	2.6
Zn^{2+}	$BHEG^-$	4.5	5.0

Abbreviations:

EDTA = Ethylenediamine tetraacetate
BHEG = N,N-bis-(2-hydroxyethyl)-glycine

factor is 8.6 while the masking factor is only 5.7. This case is an illustration that the selectivity and masking factors enable the prediction of whether a given masking agent is suitable or not for the masking of a given metal ion in a given reaction. Of course, determination of these factors requires the knowledge of the corresponding [M] values. These can be calculated with the aid of the conditional equilibrium constants. If only the thermodynamic equilibrium constants are available, they must be corrected with the factor α, as has been discussed in the preceding section.

Selectivity index

Belcher has introduced the concept of the selectivity index[40, 43] to denote the extent of the selectivity or specificity of various analytical reactions. By means of this one can easily specify the extent of selectivity of a given reaction in the course of a given analytical determination under various conditions. The symbolism of the selectivity index resembles that of an isotope; for instance, the selectivity index for the gravimetric determination of palladium with dimethylglyoxime has the following form:

$$^{\alpha}_{1}G^{Pd}$$

where α indicates that the reaction is specific to a single metal ion, Pd shows that the reaction is utilized for the determination of palladium, G stands for the gravimetric method and the subscript 1 at the left means that the determination is carried out at pH 1. Any necessary masking agent may be denoted by a subscript at the right. For example, the selectivity index of the combined extraction-spectrophotometric determination of nickel with dimethylglyoxime is as follows:

$$\begin{array}{cc} \alpha & Ni \\ & Sx \\ 7\text{-}10 & NH_3 \end{array}$$

The reaction is specific for nickel ions in ammoniacal medium at pH 7–10. All this information is unequivocally coded in the above formula.

As appears from these formulas, Belcher has designated the extent of selectivity of reactions by Greek letters. On the basis of their selectivity, he classified reactions in five groups:

The number of ions which give the reaction	Designation of the group
1	α
2–3	β
4–6	γ
7–10	δ
>10	ε

The methods applied are denoted by the following letters: G = gravimetry; S = spectrophotometry; T = volumetric analysis; Sx = extraction followed by spectrophotometry; F = fluorimetry; N = nephelometry; Q = qualitative detection; I = indicator and M = masking agent.

Table 13 lists the most important masking agents. Some of them, however, which have proved particularly useful in practice, are now surveyed in some detail.

Aminopolycarboxylic acids (EDTA, NTA) are useful for the masking of multivalent metals in neutral and alkaline media. If the amount of metal to be masked is known, the minimum necessary amount of the masking ligand is easily calculated by means of the conditional equilibrium constants. The calculation is made simple by the fact that these ligands form chelates of composition metal:ligand 1:1, i.e. no complication arises with stepwise complex formation. Demasking too is accomplished without difficulty, as aminopolycarboxylic acids are precipitated by strong mineral acids and are oxidized by permanganate.

The application of *ethylenediaminetetraacetic acid* can enhance the selectivity of inorganic precipitating agents such as halides, sulfate, tellurite, etc.[115] Only a few metals precipitate in the hydroxide form in the presence of EDTA. EDTA also increases the selectivity of sulfide precipitation.[168]

In addition, EDTA considerably increases the selectivity of many organic reagents, such as diethyldithiocarbamate, 8-hydroxyquinoline, dithizone, 2-thenoyltrifluoroacetone, etc.

Tartaric and citric acids are suitable for the masking of tri- and higher-valent metal ions in alkaline medium. For instance, they prevent the hydrolysis of iron(III), chromium(III), titanium(III), etc., in alkaline medium.

TABLE 13. MASKING AGENTS

Metal	Masking agents
Ag	CN^-, NH_3, $S_2O_3^{2-}$, Br^-, I^-, Cl^-
Al	F^-, $C_2O_4^{2-}$, OAc^-, cit, tart, EDTA, OH^-, BAL, NTE
As	S_2^{2-}, OH^-, BAL
Au	CN^-, Br^-, $S_2O_3^{2-}$
Ba	EDTA, cit, tart, NTA, DHG, SO_4^{2-}
Be	F^-, cit, tart
Bi	cit, tart, EDTA, I^-, Cl^-, NTA, DHG, TU, NTE
Ca	EDTA, cit, tart, $P_2O_7^{4-}$, NTA, DHG
Cd	EDTA, CN^-, $S_2O_3^{2-}$, SCN^-, I^-, cit, tart, NTA, DHG
Ce	F^-, NTA, EDTA, DHG, cit, tart, thyronine
Co	NH_3, NO_2^-, SCN^-, CN^-, H_2O_2, $S_2O_3^{2-}$, NTA, EDTA, DHG, cit, tart, En, BAL, *tren*, *penten*
Cr	EDTA, NTA, cit, tart, NTE
Cu	NH_3, I^-, SCN^-, CN^-, $S_2O_3^{2-}$, TU, EDTA, S^{2-}, DTC, DHG, BAL, cit, tart, NTA, TG, NTE, *tren*, *penten*
Fe	F^-, PO_4^{3-}, $P_2O_7^{4-}$, NTA, EDTA, DHG, DTC, cit, tart, BAL, SCN, $S_2O_3^{2-}$, $C_2O_4^{2-}$, CN^-, thyronine, ACB, TU, PH, S^{2-}, TG, NTE
Ge	$C_2O_4^{2-}$, F^-
Hf	F^-, SO_4^{2-}, DHG, cit, tart, NTA, EDTA, $P_2O_7^{4-}$, PO_4^{3-}, $C_2O_4^{2-}$, NTE
Hg	I^-, SO_4^{2-}, CN^-, NTA, EDTA, DHG, Cl^-, cit, tart, NTE, *tren*, *penten*
Ir	SCN^-, cit, tart, TU
Mg	NTA, DHG, EDTA, $C_2O_4^{2-}$, cit, tart, OH^-, $P_2O_7^{4-}$
Mn	F^-, $C_2O_4^{2-}$, $P_2O_7^{4-}$, NTA, EDTA, DHG, cit, tart, BAL, NTE, oxidizing agents
Mo	SCN^-, $C_2O_4^{2-}$, H_2O_2, cit, tart, EDTA, NTA, thyronine
Nb	F^-, OH^-, cit, tart, $C_2O_4^{2-}$, H_2O_2, thyronine
Ni	CN^-, SCN^-, NTA, EDTA, NH_3, cit, tart, *tren*, *penten*
Os	CN^-, SCN^-
Pb	I^-, OAc^-, $S_2O_3^{2-}$, cit, tart, NTA, EDTA, DHG, SO_4^{2-}
Pd	CN^-, I^-, NH_3, NO_2^-, SCN^-, $S_2O_3^{2-}$, NTA, EDTA, DHG, cit, tart, NTE
Pt	NH_3, I^-, CN^-, NO_2^-, SCN^-, $S_2O_3^{2-}$, NTA, EDTA, DHG, cit, tart
Rh	TU, cit, tart
Sb	cit, tart, BAL, I^-, S^{2-}, OH^-, F^-
Sn	cit, tart, BAL, I^-, $C_2O_4^{2-}$, OH^-, NTE
Sr	SO_4^{2-}, NTA, EDTA, DHG, cit, tart
Ta	F^-, OH^-, cit, tart
Th	F^-, NTA, EDTA, DHG, cit, tart, NTE, OAc^-, SO_4^{2-}
Ti	SO_4^{2-}, OH^-, H_2O_2, F^-, NTA, EDTA, cit, tart, thyronine, NTE
Tl	Cl^-, CN^-, NTA, EDTA, cit, tart, NTE
U	F^-, CO_3^{2-}, $C_2O_4^{2-}$, H_2O_2, cit, tart
V	CN^-, H_2O_2, F^-, EDTA, thyronine, NTE
W	thyronine, F^-, SCN^-, H_2O_2
Zn	CN^-, NTA, EDTA, DHG, cit, tart, SCN^-, BAL, OH^-, NH_3, *tren*, *penten*
Zr	F^-, SO_4^{2-}, cit, tart, NTA, EDTA, DHG, H_2O_2, $P_2O_7^{4-}$, PO_4^{3-}, $C_2O_4^{2-}$, NTE

Abbreviations: $C_2O_4^{2-}$ = oxalate, OAc^- = acetate; cit = citrate; tart = tartrate; BAL = 2,3-dimercapto-1-propanol; DHG = *N,N*-(2-hydroxyethyl)-glycine; NTA = nitrilotriacetate; En = ethylenediamine; TU = thiourea; PH = *o*-phenanthroline; TG = thioglycollic acid mercaptoacetic acid; NTE = nitrilotriethanol; ACB = ascorbic acid; DTC = diethyldithiocarbamate; *tren* = triaminotriethylamine; *penten* = tetraaminotetraethylenediamine; DAB = diaminobenzidine.

Triethanolamine is capable of masking aluminium(III), iron(III), manganese(II), and other ions. Other polyamines are also well applicable for the keeping in solution or masking of transition metal ions, while the alkaline earth metals remain free under such circumstances.

The *cyanide ion* is useful for the masking of elements of groups IB, IIB and VIII. Due to its extreme toxicity, hydrogen cyanide is utilized only in neutral or alkaline solutions. Decomposition of excess cyanide and demasking of some low-stability cyanide complexes (cadmium, zinc) are carried out with formaldehyde.

The *fluoride ion* forms especially stable complexes with beryllium, aluminium, iron(III), germanium, boron, tin, antimony and rare earth metal ions, etc. On heating with strong mineral acid, fluoride complexes are demasked.

Some more important methods of *demasking*:

Displacement. The masking ligand can be detached from the masked metal ion by means of another metal ion which forms a more stable complex than the original masked ion. The equilibrium may be shifted by adding a great excess of the replacing metal ion even if the stability of the new complex is not higher than that of the original masked complex. For instance, Přibil[361] masked Co(II) ions with EDTA and demasked by addition of excess Ca(II) ions. The released cobalt(II) ions were then determined with diethyldithiocarbamate which does not react with calcium. Přibil and Burger[360] demasked thorium from its EDTA complex with a great excess of sulfate. Demasking of fluoride-masked metal ions is usually achieved with excess of beryllium or aluminium ions.

Acidification. The stability of metal complexes formed with anions of weak acids depends greatly upon the pH of the solution, decreasing in parallel with the pH. For example, barium is not precipitated by sulfate from alkaline solutions containing EDTA. However, at pH 5 or lower the EDTA complex of barium decomposes and barium sulfate precipitates.

Oxidation or reduction. There are frequently considerable differences between the stabilities of complexes formed by some masking agents with the various oxidation states of a metal. In such cases demasking can easily be accomplished by oxidation or reduction of the central atom. For instance, copper(I) forms a highly stable complex with thiosulfate whereas the stability of the thiosulfate complex of copper(II) is low. Thus, by oxidation of the copper(I) complex the metal is easily demasked.

Volatility. Ligands which are readily transformed to volatile compounds, e.g. the fluoride and cyanide ions which are converted by strong acids to volatile hydrogen fluoride and hydrogen cyanide, respectively, can be removed from acidic solution by heating. Silicon and boron can be distilled as fluorides.

II. EFFECT OF CHANGES IN THE ELECTRONIC STRUCTURE OF THE CENTRAL ATOM UPON THE SELECTIVITY OF THE REACTION

In the discussion of the stability relations it was pointed out that changes in the electronic structure of the central metal atom affect the complex stability. At the same time, such changes also modify the magnetic properties, spectrum and reactivity of a complex. Although changes in the ultraviolet and visible spectra may affect the analytical applicability of the complex, reactivity changes (the possibility of mixed coordination of ligands, replacement of ligands, etc.) are of direct analytical significance.

Hence it appears even from this short introduction that if complexation with a given ligand is accompanied by large differences in the extent of distortion of the electronic structure from metal to metal, then this type of reaction will possess considerable analytical selectivity. There are several possible reasons:

1. The ligand may act as a strong-field ligand towards some transition metals and as a weak-field ligand towards others.[330] In the strong-field case complexation is accompanied by pairing of the unpaired d-electrons of the central atom. Both the structure and the absorption spectrum of the resulting low-spin complex will differ from those of the high-spin complexes.

2. In complexes of transition metals with ligands of π-acceptor ability there is a possibility of metal $\rightarrow$ ligand electron transfer of a varying extent, depending upon the energy relations and symmetry of the metal d-electron orbitals and the π-orbitals of the ligand.[84, 113, 154, 459, 485] The donor π-bond may result in a characteristic absorption band in the spectrum of the complex.

The pairing of d-electrons or the establishment of a donor π-bond as an effect of a strong-field ligand may significantly influence the ratio of stability constants of the complex.

Evaluation of either of these phenomena requires a knowledge of some fundamental bond-theory relationships.

1. LIGAND-FIELD EFFECT

In free transition metal ions (i.e. in the gaseous state) the five d electronic orbitals (d_{xy}, d_{xz}, d_{yz}, $d_{x^2-y^2}$ and d_{z^2}) are energetically equivalent. However, in the course of complexation, as a result of the electrostatic effect of ligands upon the central atom, the five d orbitals become non-equivalent. As a result of the electrostatic repulsion of the ligand, the energies of those orbitals which point at ligands will be higher than those of orbitals more remote from the negative charge of the ligands. Figure 6 shows the spatial arrangement of the five d orbitals. Which of these will be of higher and lower energies depends upon the symmetry of the complex (the steric arrangement of the ligands). For instance, in octahedral complexes the ligands lie in positions on the x, y and z axes. Inspection of Fig. 6 reveals that the lobes of only the $d_{x^2-y^2}$ and d_{z^2} orbitals point directly at ligands. Thus the energies of these will be relatively high, while those of the d_{xy}, d_{xz} and d_{yz} orbitals which point between the ligands will be lower. If the symmetry of the complex is regular octahedral, i.e. each ligand is at the same distance from the central atom, the energies of the $d_{x^2-y^2}$ and d_{z^2} orbitals will be the same (these two orbitals of identical type are denoted by e_g) and similarly the energies of the d_{xy}, d_{yz} and d_{xz} orbitals will be the same (these are the t_{2g} orbitals). Consequently, the fivefold degenerate d shell of the free ion splits under the influence of the crystal-field to doubly and triply degenerate sub-shells. Distortion of the octahedral symmetry causes further splitting of the sub-shells.

In *tetrahedral* complexes the ligands fall between the x, y and z axes as illustrated in Fig. 6. In this case the d_{xy}, d_{xz} and d_{yz} lobes point towards the ligands, and thus these are the higher-energy orbitals, while the $d_{x^2-y^2}$ and d_{z^2} are the lower-energy

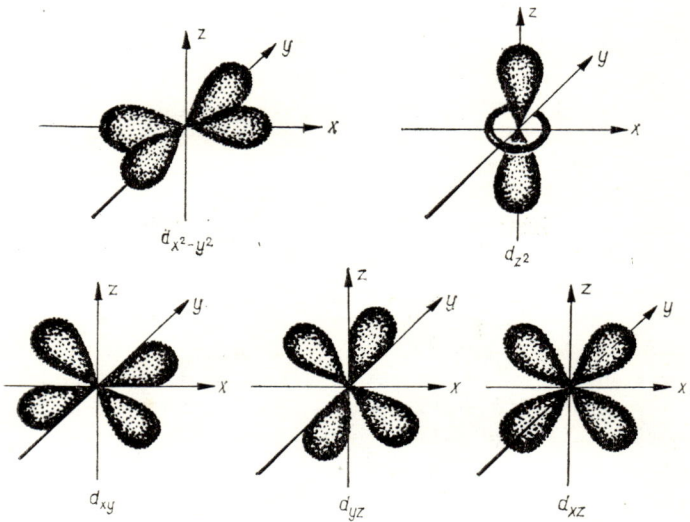

FIG. 6. Spatial arrangement of the d-electron orbitals

orbitals. Figure 7 shows the *energy-level diagram* of some complexes with various symmetry. From Fig. 6 and the above considerations, the energy sequence of the orbitals is obvious. The energy difference arising as an effect of the crystal-field and marked by Δ in Fig. 7 is termed the crystal-field splitting.

In accordance with Hund's multiplicity rule, the electrons tend to have their spins oriented parallel. Electron pairing requires the investment of the electron

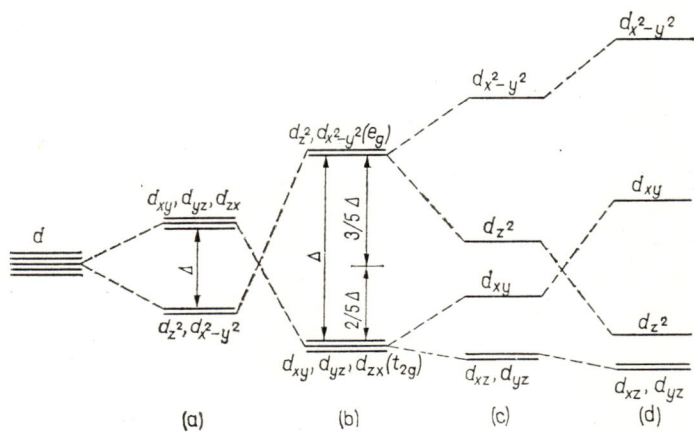

FIG. 7. Energy levels of d-electrons in (a) tetrahedral; (b) regular octahedral; (c) tetragonal bipyramidal; and (d) square planar complexes

spin pairing energy. This rule naturally holds true for both transition metal ions and complexes. In octahedral complexes, one, two or three d electrons occupy the lower-energy, and therefore more stable, t_{2g} orbitals with parallel spin. Eight, nine or ten d electrons also can occupy the orbitals in one manner only; in each case six electrons occupy the more stable t_{2g} orbitals with paired (anti-parallel) spin and the remaining electrons occupy the e_g orbitals in accordance with the multiplicity rule. However, four, five, six and seven d electrons can each occupy the d orbitals in two different manners, depending on whether the crystal-field splitting (Δ) or the spin-pairing energy (P) is the greater. Hund's multiplicity rule holds true only if $\Delta < P$. In this particular case a *high-spin* electronic structure is obtained. In complexes of strong-field ligands, $\Delta > P$. In such cases the fourth and fifth electrons, for instance, undergo pairing in the t_{2g} shell instead of occupying the e_g shell. Such an electron arrangement is termed the *low-spin* electronic structure. Hence, this latter case, when there is a chance of the formation of complexes with two different electronic structures, may also be utilizable from an analytical point of view.

With metal complexes the electrostatic field of the ligand provides the crystal-field which splits the d orbitals. However, the field-strength of ligands is influenced not only by electrostatic interactions but by many other factors (the extent of covalency of the metal-ligand bond, donor or acceptor π-bonds, etc.). The low- or high-spin electronic structure of the central atom in a complex depends upon whether the ligand-field splitting energy or the electron-pairing energy is the greater.

Fajans and Tsuchida[154, 205, 334, 459, 460] observed that replacement of a ligand by another one in a transition metal complex is always accompanied by a shift of the absorption band in a given direction within the visible region of the complex, independent of the metal. On the basis of this observation, ligands were arranged in a *spectrochemical series*. Later, the spectrochemical series was found to correspond to the order of increasing ligand-field splitting of the ligands:

$$I^- < Br^- < CO_3^{2-} < SCN^- \approx Cl^- < NO_3^- < F^- \approx (C_6H_5)_3PO \approx N_3^-$$
$$< \text{urea} \approx OH^- \approx HCOO^- < C_2O_4^{2-} \approx H_2O < NCS^- < \text{glycine} \approx EDTA$$
$$< \text{pyridine} < NH_3 < \text{ethylenediamine} < 2,2'\text{-dipyridyl} < 1,10\text{-phenanthroline}$$
$$\ll CN^-$$

From spectrometric examinations, additional ligands can be inserted in this series, e.g. diethyldithiophosphate between Cl^- and F^-, diethylenetetraamine beside ethylenediamine, dimethylglyoxime before 2,2'-dipyridyl, etc.

The high field-strengths of dipyridyl, phenanthroline and cyanide, occupying the last positions in the spectrochemical series of ligands, are a consequence of their high π-acceptor abilities, which makes it possible for them to accept electrons from the non-σ-bonding d-orbitals of the metal (i.e. to establish low-energy bonding molecular orbitals between the unoccupied π-orbitals of the ligand and filled d-orbitals of suitable symmetry of the metal; these are then occupied by the d electrons). This type of bond is termed a dative or donor π-bond, while the phenomenon itself is back-coordination. Back-coordination increases the field-strength of ligands within a complex molecule and therefore facilitates the formation of the low spin electronic structure of the central atom.

In those ligands which stand at the beginning of the spectrochemical series of ligands, i.e. which are of low field-strength, the π-orbitals are generally occupied. Hence these are bonded to the central atom by coordinative σ-bonds only, or eventually by π-bonds associated with a ligand $\rightarrow$ metal electron transfer.

The decreasing order of the Δ values of halides, $F^- > Cl^- > Br^- > I^-$, is explained by the increasing polarizability of these ligands in the same order.

Considerably less is known about the ligand-field splitting values of ligands with heavy donor atoms. It is certain, however (as has already been mentioned) that due to their enhanced π-acceptor abilities these ligands stabilize the lower oxidation states of transition metal central ions. Hence they may be strong-field ligands in complexes containing low oxidation state metals, but weak-field ones in complexes of the same metals in higher oxidation states.

As appears from the above, ligand-field splitting is really a rather complicated phenomenon, the extent of which depends on the electrostatic interaction between the central ion and the ligand, back-coordination, π-bonds associated with ligand $\rightarrow$ metal electron transfer, etc.

To a practical analyst the spectrochemical series is useful mainly in connection with spectrophotometric and colorimetric analytical methods. The series provides valuable information about the position within the visible region of absorption bands of complexes of various metal ions with different ligands, and the expected effect upon the spectrum of a complex of replacement of the ligand by some other ligand. It should be mentioned that this series provides information only about those bands which are associated with d–d electron transitions (i.e. ligand-field splitting). In addition, charge-transfer bands may appear in the visible region; these will be discussed later. With complexes containing organic ligands, additional bands due to the ligands themselves may be expected within the ultraviolet and visible regions. The analytical significance of these is less. In spectrophotometric measurements the quantitative formation of the complex in question must be ensured with an excess of the ligand, and subsequent measurement of the absorption of the complex should be performed only in spectral regions where the bands of the ligand itself are absent.

2. EFFECT OF BACK-COORDINATION

In complexes of transition-metal ions with ligands possessing free π-electron orbitals, we must always expect the establishment of back-coordination (donor π-bond associated with a metal $\rightarrow$ ligand charge transfer) in addition to the usual ligand $\rightarrow$ metal coordinative σ-bond.[87]

Depending upon the energy relations of the orbitals, there is a possibility of metal $\rightarrow$ ligand electron transfer from those lobes of the metal d-electron orbitals which overlap with the free π-orbitals of the ligand. For instance, with octahedral complexes, the t_{2g} are non-σ-bonding orbitals, while in tetrahedral complexes each d-orbital may participate in the establishment of donor π-bonds.

The orbitals of the ligand which participate in donor π-bonds may be p_π-orbitals, e.g. in chloride, bromide and iodide ions; d_π-orbitals, as in phosphine, arsine, or sulfur compounds; and delocalized π molecular orbitals, as in unsaturated hydrocarbons.

Whenever the energy relations of the corresponding d-orbitals of the metal and the free π-orbitals of the ligand make possible the establishment of a donor π-bond, the stability of the t_{2g} orbitals and the extent of ligand-field splitting increase. At the same time, the metal → ligand electron transfer increases the electron density on the ligand and hence its basicity, while it decreases the negative charge density on the metal atom and hence increases its electrophilic character. This double effect strengthens the coordinative σ-bond. On the other hand, by increasing the negative charge density on the central metal ion, the coordinative σ-bond increases the extent of back-coordination. The compensating effect of these two bond types upon the negative charge density on the central atom is well illustrated, e.g. by the Mössbauer studies of Burger et al.[92]

Back-coordination has fundamental importance in the formation of metal carbonyls, phosphine and arsine complexes, and metal complexes of unsaturated hydrocarbons, but due to its effect upon complex stability its study is important in all those cases when the ligand possesses π-acceptor ability.

Back-coordination is significant from the analysts' viewpoint in many respects:

1. Enhancement of the complex stability, as has already been mentioned, increases the selectivity of the reaction. Those metal complexes of a given ligand, whose central atom is capable of back-coordination, will possess outstanding stability, and will consequently be formed even at low concentrations of the free ligand or in strongly acidic medium.

The great majority of organic ligands used for analytical purposes are conjugate bases of acids. Thus the concentration of the free ligand is determined in almost all cases not only by the overall concentration of the reagent but also by the pH of the solution.

2. Back-coordination may modify the ratio of stepwise equilibrium constants favourably. Thus, for example, the extremely high value of the third stepwise stability constant in the 1,10-phenanthroline complexes of Fe(II) can be interpreted by such bonding. The high stability of the MA_3 complex results in the quantitative formation of this species with even a small excess of the ligand. Since the other stepwise complexes (MA, MA_2) are then absent from the solution, the reaction is clear-cut; this is a requisite of its analytical application.

3. The energy relations of the electronic orbitals of the central atom and the ligand which lead to back-coordination also facilitate the appearance of charge transfer absorption bands. These bands are much more intense than those associated with the d–d transitions dealt with in the preceding section,[238] and therefore they make possible the elaboration of much more sensitive spectrophotometric methods. For instance, the analytically utilized red colour of the iron(II) 2,2'-dipyridyl complex originates from a transition of the electrons of the central atom to the π-orbitals of the ligand.

It should be mentioned that charge-transfer bands are produced not only by electron transfer from the central atom to orbitals of the ligand but also by excitation of electrons in the opposite direction. For example, the red colour of iron(III) thiocyanate originates from excitation of the electrons of the thiocyanate to orbitals of the iron(III) ion. In general, it can be stated that the appearance of charge-transfer bands associated with ligand → metal electron transfer is expected in the spectrum of a complex consisting of a high oxidation state central atom and a reducing ligand; in the case of a complex formed by some low oxidation state central atom with a π-acceptor ligand the bands associated with metal → ligand electron transfer will be observed. Back-coordination is of importance in this latter type of complex.

III. ANALYTICAL SELECTIVITY OF THE FORMATION OF MIXED COMPLEXES

From an analytical point of view, the most promising specific reactions are those based on the formation of mixed ligand complexes. In numerous chelate complexes the metal is coordinatively unsaturated and is capable of the further coordination of monofunctional ligands of relatively small volume. Formation of a mixed complex modifies the electronic structure of the central atom, as well as the charge and symmetry of the complex molecule. Consequently, the physical and chemical properties of the mixed complex will be different from those of the parent complex. As regards analytical application, the most important factor is the effect of the formation of the mixed complex upon the solubility and the absorption spectrum. We shall return later to a brief discussion of the factors determining the solubility and light absorption of a mixed complex.

Conditions for the analytically selective formation of mixed complexes

Whenever a metal ion exists in solutions together with two or more different ligands, there is always the possibility of the formation of mixed ligand complexes. The actual formation of a mixed complex depends on the affinities of the metal ion towards the various ligands present and the relative concentrations of these ligands.

In general, the analytically important mixed complexes contain both chelating ligands and monodentate ligands. The chelating ligands are usually bound more strongly than the monodentate ligand to the central atom. The main requirements for the formation of this type of mixed complexes are the following: (i) the chelating ligand should be sterically unable to occupy each of the coordination sites of the central atom (i.e. to coordinatively saturate the central atom); (ii) the monodentate ligand should be small enough to be able to occupy the free sites in the coordination sphere of the central atom; (iii) the concentration of the monodentate ligand should be sufficiently high. The formation of such mixed complexes also depends on the electronic structure and ionic radius of the central metal. The monodentate ligand can coordinate only if there is an orbital in the electronic shell of the central atom to accept the unshared electron pair of the ligand and if the radius of the central atom is sufficiently great as to allow the approach of the ligand. In addition to these most important requirements, the mutual polarizability of the central atom and the ligands, the π-acceptor ability of the ligand, and the symmetry relations of the electronic orbitals of the central atom and the ligands affect the formation of the mixed complex considerably.

From the above considerations one can easily understand why only a fraction of the many metal complexes of a given chelating ligand are capable of the coordination of further monofunctional ligands.

The analytical selectivity of the formation of mixed complexes is excellently illustrated by bis-(dimethylglyoximato)diiodocobaltate(II).[88] Cobalt(II) ions form complexes of coordination number four and six. The preferred symmetry of the former is tetrahedral, while that of the latter is octahedral. The two nitrogen donor atoms of a dimethylglyoxime ligand occupy two of the coordination sites of the cobalt(II) ion in the cobalt dimethylglyoxime complex. The two dimethylglyoxime ligands are kept in a planar configuration by strong hydrogen bonds, and hence the tetrahedral arrangement of the donor atoms is prevented (Fig. 8).

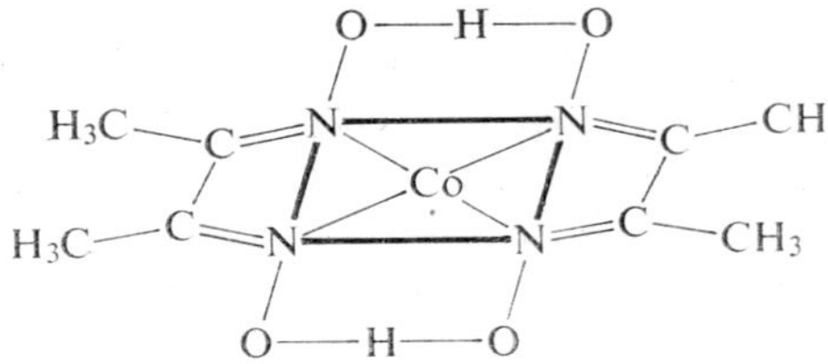

FIG. 8. Structural formula of cobalt(II) dimethylglyoxime

Thus the complex possesses a square-planar structure with respect to dimethylglyoxime. This hydrogen-bonded structure prevents the coordination of a third chelating ligand which would result in the establishment of the octahedral symmetry preferred by the cobalt(II) complex. Consequently, coordination of two monodentate ligands along the z axis of the square-planar complex is the only chance for the establishment of a complex of distorted octahedral symmetry.[93]

The divalent ions of the other $3d^5$–$3d^{10}$ transition metals also form square-planar complexes with dimethylglyoxime. Of these, for the low spin nickel(II) complex square-planar symmetry is the most suitable. Therefore this complex is the least capable of the coordination of monofunctional ligands. All the other dimethylglyoxime complexes coordinate the hydroxide ion; this is a strong base (nucleophile) of relatively small volume and therefore has great affinity for the central atom.[94] The considerably less basic and much larger iodide ion is bound (at iodide concentrations up to 0.5 M) by the cobalt(II) complex only. Hence the formation of the bis-(dimethylglyoximato)diiodocobaltate(II) mixed complex is an analytical reaction specific for cobalt. The absorption of the mixed complex at 435 nm is suitable for the selective determination of cobalt.[96]

The analytical selectivity of the formation of mixed complexes can be further enhanced by making use of the different physical and chemical properties of the parent complexes formed with the two ligands separately. This may be illustrated by the above example of the cobalt(II) mixed complex. As the mixed complex is readily soluble in water, it can easily be separated from those cations which give an insoluble precipitate with either iodide or dimethylglyoxime. Thus, cobalt(II) may be determined by the above procedure in the presence of palladium, nickel, silver, etc. The formation of the mixed complex is quantitative at about pH 6. Some metal ions undergo hydrolysis under such conditions. The water-soluble mixed complex can be extracted quantitatively from the insoluble basic salts formed by hydrolysis. Naturally, no interference is exerted by those metal ions which do not form dimethylglyoxime or iodo complexes or other species which absorb at the same wavelength as the cobalt(II) mixed complex.

The rates of formation too of mixed complexes may differ considerably from those of the corresponding parent complexes; this can often be utilized for analytical purposes. For instance, dithizone is considered in general not to be useful for the determination of nickel because of the relatively low rate of formation of the nickel dithizone complex. Freiser et al.[179] have pointed out that in the presence of pyridine or other nitrogen bases a mixed complex containing nickel and dithizone is formed almost immediately on mixing of the reactants. Hence the coordination of the base enhances the rate of formation of the complex of nickel with dithizone. They explained the phenomenon by assuming that coordination of the base facilitates the replacement of water molecules by dithizone in the coordination sphere of the nickel ions.

Several researchers have examined the effect of various ligands upon the rate of replacement of coordinated water molecules by other ligands. For instance, Margerum[284] pointed out that $CuOH^+$ ions react much faster with EDTA than do the aquo-copper ions. In contrast, coordination of acetate ions to the copper ion reduces the rate of reaction of the latter with EDTA. Hammes[203] demonstrated that the 1:1 complex of nickel and glycine releases a coordinated water molecule 13 times as fast as does a hexaaquo-nickel ion. The phenomenon was interpreted by these authors in terms of the reduced net positive charge of the central atom under the effect of the ligand of higher donor strength as compared to water, which results in a decrease of the attractive force of the central atom towards the coordinated water molecules.

Freiser et al.[176] utilized the accelerating effect of the coordination of acetate ions upon the rate of formation of the zinc dithizonate complex for the separation of zinc and nickel. Acetate ions enhance the rate of formation of zinc dithizonate to a much greater extent than that of nickel dithizonate. Thus zinc can be extracted quantitatively from nickel with dithizone in the presence of acetate ions.

Factors determining the solubilities of complexes

In analytical practice the solubilities of metal complexes are very important. For instance, the analytical selectivity of dimethylglyoxime arises from the low solubilities of its complexes with nickel and palladium, whereas the corresponding complexes of all of the other transition metals are quite soluble in water. The analytical selectivity of many other organic reagents is similarly a consequence of the solubility relations of their complexes.

The most common solvent is water. Because of their polar character, water molecules are interconnected by strong hydrogen bonds, i.e. they form more or less well defined associations. As a result of hydration, water-soluble materials modify the structure of water on dissolution.

Simple organic molecules containing hydrophilic groups, e.g. alcohols, aldehydes, ketones, carboxylic acids and amines undergo *hydration by hydrogen bond formation* in water. Sulfur atoms are less capable than are oxygen atoms of hydrogen bond formation. This is why replacement of an oxygen atom in an organic molecule by sulfur is accompanied by a decrease in water-solubility. The hydrogen bond forming ability of nitrogen is about the same as that of oxygen in analogous compounds.

Aliphatic hydrocarbon chains and aromatic rings, which cannot effect hydration in the above manner, reduce the solubilities of organic molecules in water. An increase in the magnitude of the hydrophobic character of a molecule, e.g. more hydrophobic rings, results in a reduction of the solubility in water.

The electric charge of a molecule is also of importance in determining the solubility. Charged species, either cations or anions, possess a sufficiently great electrostatic field to orientate water molecules around themselves. This explains why electrically neutral water-insoluble molecules which form anions in alkaline medium become soluble under such circumstances. Similarly, those molecules which are protonated in acidic medium to give cations are soluble in aqueous solutions at low pH. Such chemical properties of organic reagents determine in which pH region the reagents can be applied. Strongly polar materials behave somewhat similarly to ions and also undergo hydration relatively easily.

These qualitative relationships also provide information concerning the solubilities of metal complexes in water. Those organic reagents which form electrically neutral complexes with metal ions are generally useful precipitating agents in gravimetry. Exceptions are those ligands which are strongly polar and whose hydrophobicity is less than the hydrophilicity. Their complexes are quite soluble in water even if they are electrically neutral. This explains, for instance, the solubility in water of the Cu(II) glycine complex. The copper complexes of anthranilic and quinaldinic acids, containing the same donor atoms, are insoluble in water.

With metal complexes, the *association of complex molecules* (formation of polynuclear complexes) may also decrease the solubility. For instance, in the complexes of pyrogallol, for steric reasons only two of the phenolic hydroxy oxygen atoms can bind to the same metal atom, the third one is linked to the central metal atom of another, neighbouring, complex molecule.

Thus, the complexes are actually associations of an extremely large number of molecules. Their solubilities are much less than those of the analogous pyrocatechol complexes. Similar phenomena occur with the complexes of many inorganic anions. For example, carbonates, phosphates and sulfates usually give complexes in which their two donor atoms bind to different metal atoms and so the salts possess a quasi-polymeric structure.

The variously positioned *substituents* of an organic ligand can influence the solubilities of its complexes in several ways. The *steric hindrance* of the substituent, for instance, may prevent the attachment to the central atom of the number of ligands required to neutralize the positive charge of the central atom. Thus, due to the steric hindrance of the methyl group, only two molecules of 2-methyl-8-hydroxyquinoline can bind simultaneously to aluminium(III) because of its small ionic radius. Hence a positive charge remains on the complex molecule and this makes it soluble in water. This is why the reagent is applicable for the selective determination of zinc in the presence of aluminium.

If it is desired to increase the solubility in water of a chargeless metal complex, some ionizing group, e.g. sulfonic acid or carboxylic acid group, must be built into the ligand. The *hydration ability of this substituent* favourably affects the solubility of the complex. For instance, the 8-hydroxyquinoline complex of iron(III) is uncharged and therefore it is practically insoluble in water; the analogous 8-hydroxyquinoline-5-sulfonic acid complex is readily soluble in water. The diethyldithiocarbamate complex of copper is insoluble in water, but the corresponding 3-carboxy derivative is readily soluble.

A substituent which is protonated in acidic media to become positively charged increases the solubility of the complex in acidic solution. For example, introduction of a dimethylamino group in the *p*-position of phenylfluorine makes its metal complexes soluble in acidic aqueous media.[247]

With *coordinatively unsaturated complexes* coordination of negatively charged ligands to the uncharged parent complex will result in complex anions which may be readily soluble in water in contrast to the sparing or low solubility of the parent complex.

However, although the solubilities of coordinatively unsaturated complexes are affected by the attachment of added ligands to the central atom, water molecules also possess a high coordinating power. Complexes with unfilled coordination spheres are more soluble in water than filled coordination sphere complexes, since the former are capable of the coordination of water molecules. In this way the hydrophilicity and water-solubility of a complex increase together.

The effect of the formation of a mixed ligand complex upon the solubility is clearly illustrated by the case of palladium dimethylglyoxime. This is practically insoluble in acidic and neutral aqueous media, but in alkaline medium the palladium central atom can bind a hydroxide ion[89] and the resulting negatively charged complex molecule is readily soluble in water. The analogous nickel complex is incapable of the coordination of hydroxide ion even in strongly alkaline medium[268] and this is why nickel and palladium ions can be separated with dimethylglyoxime or, in other words, why dimethylglyoxime is specific for palladium(II) ions.

The differences in solubility of dimethylglyoxime complexes can also be explained in terms of the unsaturation of the coordination spheres of the central atoms.[88] Of the dimethylglyoxime complexes of the divalent $3d^5$–$3d^{10}$ transition metal ions, only the nickel(II) complex is incapable of binding monodentate ligands (water, hydroxide ions, halide ions). This is the reason for the low solubility of the nickel dimethylglyoxime in water. In the absence of monodentate ligands the other complexes bind water molecules, or hydroxide ions in alkaline medium, and this results in their enhanced solubilities in neutral and alkaline aqueous media.[94] As has already been mentioned, the two dimethylglyoxime ligands assume a square-planar arrangement in these complexes and hence coordination of two small monodentate ligands may occur along the *z* axis.

Absorption of complexes

In analytical chemistry, the colour difference of the metal complex and the free ligand is often utilized for both qualitative and quantitative analytical purposes. The colour of a molecule arises from a selective absorption by the molecule of only a fraction of the light falling on it. The light absorbed is of definite wavelengths. The absorbed energy leads to the excitation of certain electrons within the molecule. The difference between the electronic energy levels in the molecule determines the wavelength at which the material absorbs, i.e. the colour of the material.

The band appearing in the spectra of those transition metal complexes which occur most frequently in analytical practice are associated with three fundamental types of electronic transition, and accordingly can be classified into the following three groups: 1. d–d transition bands; 2. charge-transfer bands; 3. bands corresponding to transitions within the ligand.

1. In the case of a *d–d transition*, the electron is excited from one of the d orbitals of a central atom to another d orbital of the same atom, e.g. a $d_{t_{2g}}$–d_{e_g} transition. Actually, such transitions are *forbidden* and only the perturbation effects of the ligands make them possible. Due to the small difference in energy between the t_{2g} and e_g orbitals in transition metal ions (aquo complexes), the d–d bands appear in the visible region. In general, d–d transitions are responsible for the colours of transition metal ions. Replacement of water molecules in aquo complexes by some other, more basic (more strongly coordinated) ligand is associated with an increase in the difference between the energy levels of the d orbitals, the effect being caused by the stronger field of the new ligand, and thus the d–d band is shifted towards lower wavelength. If the coordination of the new ligand also modifies the geometry of the complex, the intensity of the bands increases. For instance, the molar extinction coefficient of the blue complex [$CoCl_4^-$] of tetrahedral symmetry is considerably greater than that of the pink aquo complex [$Co(H_2O)_6^{3+}$] of octahedral symmetry.

2. A *charge-transfer band* is associated with the transition of an electron from an orbital belonging mainly to a ligand to another orbital belonging mainly to the central atom, or conversely from an orbital belonging to the central atom to a ligand orbital. Charge-transfer of the first type may occur with an oxidizable ligand and a high oxidation state central atom. The more powerful oxidizing and reducing agents are the central atom and the ligand, respectively, the lower the energy required for the excitation, i.e. the charge-transfer band appears at longer wavelength in the spectrum. The charge-transfer bands of the mixed complexes [$Co(NH_3)_5X$]$^{2+}$, where X = F^-, Cl^-, Br^- or I^-, appear at wavelengths increasing in the order $F^- < Cl^- < Br^- < I^-$. The molar extinction coefficients of charge-transfer bands are about a hundred times those of d–d transition absorption bands.

If the absorption bands of the ligand itself are neglected, the charge-transfer bands of transition metal complexes appear in general in the ultraviolet region of the electron excitation spectrum, while the considerably less intense bands associated with d–d transition lie in the visible region. If the energy of a charge-transfer electron transition is sufficiently low, the charge transfer band may be shifted to the visible region where, because of its higher extinction coefficient, it may shield the d–d transition band. For instance, the vivid red colour of iron(III) thiocyanate is due to charge transfer.

Charge-transfer bands caused by the excitation of an electron of the central metal atom to a ligand orbital have already been dealt with in the section on back-coordination. Such charge-transfer bands appear in the spectra of complexes formed by low oxidation state metals with unsaturated ligands. Electron transitions of this type are significant primarily for those complexes in which there is a possibility for back-coordination associated with a metal-to-ligand electron transition even without excitation radiation. Such a charge-transfer band is responsible for the red colour of the iron(II) 2,2′-dipyridyl complex well known in analytical chemistry.

3. In a survey of the spectra of complexes formed with organic ligands, mention should be made of the *bands associated with electron transitions within the ligand*. In many cases, the spectrum of a complex is very similar to that of the free ligand (the organic molecule) because of the intensity of the bands of the ligand.

In those coordination compounds where the central atom is linked to the ligand primarily by electrostatic forces, the metal ion modifies the spectrum on the ligand in much the same way as does protonation of the ligand. In such cases complex formation is accompanied by only a small displacement of the ligand bands towards shorter, or more rarely towards longer, wavelengths. The shape of the absorption curve and the values of extinction coefficients at the peak maxima are only slightly, if at all, modified. Displacement of the absorption bands is quite uncharacteristic of the nature of the cation in these compounds.

With more stable complexes, where the central atom is bound by a more or less covalent bond to the ligand, the extent of the modification in the spectrum of the ligand also depends on the degree of covalency of this bond. The absorption maxima of complexes of eriochrome black T appear at shorter wavelengths than that of the free ligands. With an increase in complex stability the absorption maxima are shifted towards the ultraviolet region.

In those complexes where the nature of the bond between central metal atom and ligand is very covalent, the characters and energies of the molecular orbitals of the ligands and accordingly the probabilities of electron transitions are greatly altered. In the spectra of these complexes charge-transfer bands occur in addition to the absorptions of the ligands, particularly if complex stability is enhanced by back-coordination. In many cases, differentiation between these two types of absorption is rather difficult, especially with more complicated complexes.

Formation of *mixed ligand complexes* affects the positions of bands of all three types. The field strength is usually greater in a mixed complex than in the corresponding parent complexes. Accordingly, ligand-field splitting too is greater and the $d-d$ bands are shifted towards the ultraviolet region. The modified symmetry relations of mixed complexes, as compared to those of the corresponding parent complexes, result in changes in band intensities (extinction coefficients).

The coordination of the new ligand or ligands to the central atom in mixed complexes may lead to new charge-transfer absorption bands. Such a new charge-transfer band, for example, is that of the mixed complex bis-(dimethylglyoximato)-diiodocobaltate(II) which is used for the spectrophotometric determination of cobalt.

In solution, nickel dithizonate is capable of the coordination of various bases. Coordination of *o*-phenanthroline to the planar nickel dithizonate complex results in the appearance of a new characteristic high-intensity charge-transfer band[287, 288] which may serve as the basis of one of the most sensitive procedures for the determination of nickel.

In mixed complexes the strengths of the bonds between the different ligands and the central atom are not identical with those in the corresponding parent complexes. The presence of a second ligand may either increase or decrease the bond strength between the first ligand and the central atom. Consequently, the energies of the electron orbitals in the ligand are altered; this is accompanied by displacement of the absorption bands due to electron transitions within the ligand.

IV. FUNCTIONAL GROUPS AND ANALYTICAL SELECTIVITY

After examining the reactions of many organic molecules with various metal ions, analysts associated the selectivity of individual reagents with the presence of certain functional groups. The specificity of individual functional groups for metal ions was determined empirically. The real origin of analytical selectivity is studied today by coordination chemical methods. For certain selective reactions the reasons have already been determined, either in part or fully, but in other cases further research is required. It is quite clear from the subject matter of the previous sections that the analytical selectivity of reactions based on complexation does not imply that a ligand applied as reagent forms a complex with only one or a few metal ions. A reagent may be selective, for instance, if it gives complexes with many metal ions, but only one or two of these are insoluble in water. In another case, the ligand may form complexes of about the same stability with various metal ions, but only one of them absorbs within a certain spectral range; the reagent may then be specific for that metal ion. Similarly, other physical or chemical properties of a complex may ensure selectivity for a reagent.

This section deals with the phenomenon of analytical selectivity. Examples will be presented using available experimental data for some of the more important organic metal reagents and the conditions determining analytical selectivity and limitations of application will be specified.

In addition to the functional groups participating directly in complexation, i.e. containing the donor atoms, the selectivity of a ligand applied as an analytical reagent depends fundamentally upon those of its substituents which affect the solubility, colour and other physical properties of the complex. Thus, the analytically important groups fall into the following two classes: 1. *functional groups* containing the donor atoms which participate directly in the bond with the metal atom; 2. *substituents* which determine the other properties of the complex.

However, there is a series of other factors which play an important role in establishing the analytical selectivity of reactions. The reagents and in some cases the products also must exist in solution, and hence their solubility is an important factor. For instance, in those cases where the organic reagent is soluble in alkaline aqueous medium only, the concentration of the free ligand necessary for complex formation can be attained in the presence of alkali. It may also occur that the organic reagent exists in different forms (tautomerism) depending upon the pH of the solution, and only one of these can react with the metal ion in question. Under such circumstances it is the pH of the solution which determines whether the reaction takes place or not. Again, there are cases when the complexation reaction proceeds only in alkaline medium which ensures a sufficiently high concentration of the free ligand; however, due to its inert character the resulting complex is resistant to acids.

It was seen above that it is practically impossible to speak of specific or selective atomic groups without specifying the reaction conditions. These latter are of the same importance in establishing analytical selectivity as the functional groups of the reagent. Actually, we should speak of *specific* or *selective reactions* only, stressing the effect of the medium (pH, masking agents, solvent, etc.).

Apart from a few quite special cases, the *donor atoms* of organic reagents applied for analytical purposes are oxygen, nitrogen and sulfur.

In the functional groups of chelating ligands the *oxygen donor atoms* can be present in phenolic, carboxylate or alcoholic groups, the C=O group of aldehydes or ketones, or as ethereal oxygen.

The *nitrogen donor atom* may be in the form of primary, secondary or tertiary amines, nitro, nitroso, azo, or diazo groups and in some cases nitrile or carboxamide groups.

The *sulfur donor atom* may be present in an ionized thiol or a thiocarboxylate anion, a thioether, thioketone or a disulfide.

Apart from a few exceptions, organic reagents are chelate-forming ligands, i.e. they contain at least two donor atoms so arranged that the two or more donor atoms of each ligand can bind to the *same* metal atom to give five- or six-membered rings. The most frequently occurring chelate rings in metal complexes or organic reagents are depicted in Fig. 9.

Ring I occurs in complexes of 8-hydroxyquinoline and its derivatives, and of 4-hydroxybenzothiazole. Ring II is formed in complexes of 2,2'-dipyridyl, 1,10-phenanthroline and α,α',α''-tripyridyl. Ring III occurs in dioxime chelates; ring IV in chelates of salicylaldoxime and its derivatives; ring V in those of pyrocatechol, pyrogallol and their derivatives; ring VI in those of ethylenediamine and its derivatives; ring VII in those of *o*-aminocarboxylic acids; and ring VIII in chelates of α-hydroxycarboxylic acids.

In the following the application of some of the more important organic reagents will be outlined schematically. The reagents will be dealt with in the order of their

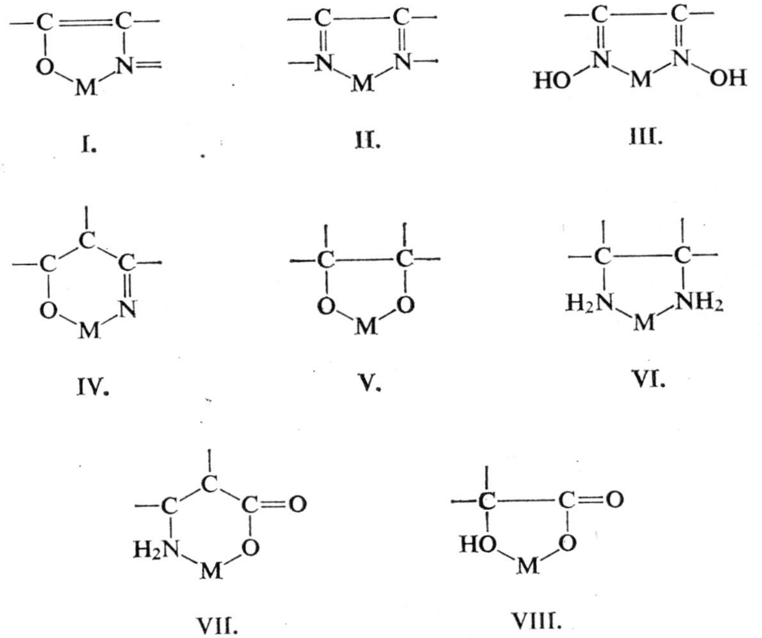

FIG. 9. Some of the most frequently occurring chelate rings in metal complexes of organic reagents

chelate rings according to Figure 9. The degree of selectivity of the individual ligands is indicated in each case on the basis of experimental data; the reason for the selectivity is dealt with only in those cases where the available results permit it.

8-Hydroxyquinoline (oxine) and its derivatives

Oxine (HA) gives chelate complexes of composition MA_2 with a series of divalent metal ions, MA_3 with trivalent metal ions, and MA_4 with thorium and zirconium; all are insoluble in water. Due to the systematic research work of Berg[49] in this field, although this compound reacts with many metal ions, its reactions can be made selective in many cases by appropriate choice of the reaction conditions. For instance, it is suitable for the selective detection of aluminium(III) or iron(III) ions in the presence of phosphate, tartrate, oxalate or malonate as masking ligands.

Sandell[396] utilized the fluorescence of the oxinates of gallium, aluminium, indium and zinc for the quantitative determination of these elements.

Berg[48] pointed out that by introduction of appropriate substituents the selectivity of oxine can be enhanced. For instance, the complexes of the dichloro and dibromo derivatives of oxine are much more resistant to strong mineral acids than those of the parent ligand.

In analytical practice 7-iodo-8-hydroxyquinoline-5-sulfonic acid[487, 488] is considered the most selective derivative of oxine. This reagent forms a vivid blue complex with iron(III) ions at about pH 2.5. Like oxine, this reagent also reacts with a great variety of metal ions, but of these only the iron(III) and vanadium complexes are coloured. Consequently, it can be utilized for the *specific* determination of iron in the presence of almost any other metal. For this reason it has been named *ferron*.

Ferron is an excellent example to illustrate the difference between analytical and complex chemical selectivity. In addition to being an almost universal complexing agent, it is analytically specific for iron. With its aid iron can be detected in dilutions as high as 1:10,000,000. Its iron complex is so stable that the ligand extracts iron from ferric phosphate.

4-Hydroxybenzothiazole

A comparison of the structure of this reagent with that of 8-hydroxyquinoline dealt with above reveals that the two are closely related. Nevertheless, their analytical behaviours are different, confirming what has already been stressed, that the selectivity of a given reagent does not depend exclusively, or even primarily, upon its functional groups.

For instance, 4-hydroxybenzothiazole does not react at all with titanium, vanadium, molybdenum or tungsten, whereas oxine is a sensitive reagent for these metals.

Nitrosonaphthol ligands

1-Nitroso-2-naphthol was the first known selective organic reagent;[198] it was used in the detection and determination of cobalt(III) and iron(III). A particularly advantageous feature is its applicability for the selective determination of cobalt in the presence of nickel. Although described as selective, it gives coloured precipitates with numerous metal ions (iron(III), chromium(III), copper(II), zirconium(IV), palladium(II)) in addition to cobalt. Accordingly, its selectivity refers to certain groups of metals only, as is the general case with such types of reagent. For example, it makes possible the detection of palladium in the presence of the other noble metals, and of cobalt in the presence of metals of the ammonium sulfide group and copper, etc. The ability to detect cobalt in the presence of heavy metals is rendered possible by the fact that due to the great stability of the cobalt complex the nitrosonaphthol reagent can react even with cobalt phosphate, while the majority of heavy metals can be masked with phosphate ions. The copper(II) ion is extremely well masked by iodide (elemental iodine is reduced by sulphite) which does not interfere with the determination of cobalt.

2-Nitroso-1-naphthol reacts analogously to 1-nitroso-2-naphthol. The complexes of these two ligands are identical in composition and, apart from the zirconium complex, in colour too. It is interesting that while the zirconium complex of 1-nitroso-2-naphthol is yellowish-green, that of the isomeric ligand is deep red.

These complexes are electrically neutral large molecules which are accordingly insoluble in water. By introduction of hydrophilic groups and retention of the =C(NO)–C(OH)= grouping "specific" for cobalt, water-soluble complexes suitable for the spectrophotometric measurement of cobalt can be prepared. The most frequently applied ligand of this type is the *nitroso-R salt* suggested by Van Klooster;[464] this is the disodium salt of *1-nitroso-2-naphthol-3,6-disulfonic acid*. The formation of its cobalt complex occurs in acetate buffered media. However, once complex formation is complete, the product is resistant even to strong mineral acids. This renders possible the selective detection of cobalt in the presence of nickel and iron which similarly give complexes with this ligand in acetate buffered media; the nickel and iron complexes decompose on acidification. The explanation of this phenomenon is connected with the inert character of the low spin cobalt(III) complex, in contrast with the unstable (labile) nature of the other two complexes. (The expression "unstable" (labile) refers in the present case to the kinetics of dissociation of the complex, i.e. it is the opposite of the term "inert"; it does not give any information on the equilibrium constant of complex formation).

Nitrosonaphthol derivatives were used for the determination of cobalt in analytical chemistry long before the oxidation state of the central atom was known. Finally in 1932 the examinations of Mayr[291] revealed that the central cobalt atom in the complex is cobalt(III). Today, modern bond-theory considerations make this fact appear trivial since ammonia, which possesses a much lower field strength than the ligand concerned, also forms a low spin cobalt(III) complex.

2,2'-Dipyridyl and 1,10-phenanthroline

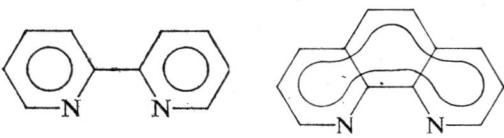

2,2'-Dipyridyl and 1,10-phenanthroline and also 2,2'-dipiperidyl were known as early as the end of the last century to give highly stable, intensely red, water-soluble complexes with iron(II). Today we know that these complexes are of low spin electronic structure (i.e. diamagnetic). The central iron(II) atom is surrounded by three ligands in octahedral symmetry. The complex formation stabilizes the iron(II) oxidation state: the formal redox-potential of the complexes (+1.06 V) is higher than that of the iron aquo complex. The stabilization of the iron(II) state is explained by back-coordination involving a metal-to-ligand electron transfer.

Back-coordination also results in the greater field-strength of the ligand. The low spin electronic structure and hence the diamagnetic nature of the MA_3 complex, and the anomalous order of the stepwise stability constants ($K_3 \gg K_1 K_2$) (which was seen in Section II to be of analytical importance) are consequences of the enhanced field-strength of the ligand.

These reagents, and also α,α',α''-tripyridyl which was suggested by Morgan,[309] are used for the determination of traces of iron.

Introduction of fairly bulky substituents on the carbon atoms adjacent to the nitrogen donor atoms destroys the capability of the reagent to form the characteristic tris complex of iron(II). In contrast, these substituents do not hinder the formation of the corresponding copper(I) complex, in which only two molecules of the ligand are attached to the central atom in tetrahedral symmetry. Consequently, such substituted dipyridyl and phenanthroline molecules act as selective reagents for copper.[222, 223, 228]

Feigl[159] utilized the iron(II) dipyridyl and phenanthroline complexes for the precipitation of some inorganic complex ions (MoO_4^{2-}, WO_4^{2-}, BiI_4^-, HgI_4^{2-}, CdI_4^{2-}, $Ni(CN)_4^{2-}$, etc.). The precipitates are various shades of red. With this reaction e.g. 0.05 γ cadmium can be detected in dilutions as high as 1:1,000,000.

Dioxximes

$$\begin{array}{c} R \\ \diagdown \\ C=N \\ | \\ C=N \\ \diagup \\ R \end{array} \begin{array}{c} O-H \\ \\ \diagdown \\ \\ O-H \end{array}$$

Dioxximes are today considered as reagents of the highest selectivity. Dimethylglyoxime is specific for nickel(II) ions in ammoniacal solution if alkali tartrate is employed as masking agent to keep in solution the possibly interfering hydrolysing salts. The palladium(II) complex of dimethylglyoxime, which is insoluble in water, is soluble in the presence of ammonia. It is noteworthy, however, that in the joint presence of iron(III) and cobalt(II) a polynuclear complex of empirical formula $FeCoC_{12}H_{19}N_6O_6$ precipitates even in the presence of tartrate. Therefore iron(III) must be reduced to iron(II) prior to the addition of dimethylglyoxime. In this case the selectivity of the formation of the nickel complex is assured in the presence of these metals, too.

The much greater stability of the palladium complex as compared to that of the nickel complex renders possible the formation of this complex and the quantitative precipitation of palladium from weak mineral acidic media. At about pH 1 the other dimethylglyoxime complexes do not form at all and hence the reaction is specific for palladium.

However, it would be a fallacy to conclude from the above statements that dimethylglyoxime reacts exclusively with the metals mentioned. There are many observations[65, 94, 112, 140, 142, 465] indicating that dimethylglyoxime forms high-stability complexes with all $3d^5$–$3d^{10}$ transition metals. For instance, its copper complex is brownish-red, its iron(II) complex is reddish, and the cobalt(II) complex is brownish. However, these chelates are rather soluble in water, and therefore the presence of these metals does not interfere with the gravimetric determination of nickel. The stabilities of these complexes are considerably lower than that of the corresponding palladium complex, and hence they do not form at all in acidic media where the palladium complex does exist (see Table 8).

The structures of dimethylglyoxime complexes are known from X-ray diffraction[56, 172–4, 194, 333, 480] and infrared spectrometric[63, 64, 97, 181, 193, 319, 320] measurements. The complexes contain the two dimethylglyoxime ligands in a square-planar configuration. The two ligands are interconnected by strong intramolecular hydrogen bonds which thus stabilize the 1:2 metal:ligand chelate. This causes the anomalous ratio of the stepwise stability constants ($K_1/K_2 \leq 1$), which makes possible the quantitative formation of the MA_2 complex with even a small excess of the reagent. X-ray diffraction data indicate that in the cases of the nickel, palladium, platinum and gold complexes the complex molecules are interconnected by metal–metal bonds in the solid state (Fig. 10), while in the case of the copper(III) complex the molecules are interconnected by metal–oxygen bonds (Fig. 11).

Infrared spectroscopic studies did not only provide information on the structures of the complexes (mainly the existence of strong hydrogen bonds); they also proved[97] the important role of back-coordination in stabilizing the complexes. The strong ligand-field of dimethylglyoxime stems from back-coordination and it is

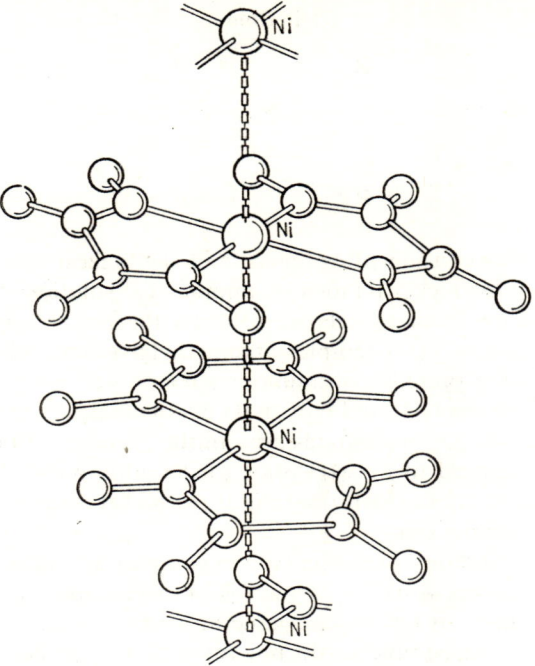

Fig. 10. Structural model of the nickel dimethylglyoxime complex. The nickel–nickel bonds can be seen

this strong ligand-field which is responsible for the low spin electronic structure of the iron(II), cobalt(II), nickel(II) and palladium(II) complexes.[92]

In connection with the determination of nickel in the presence of palladium(II) it was necessary to ascertain why the dimethylglyoxime complex of palladium dissolves in alkaline medium, while the corresponding nickel complex does not.

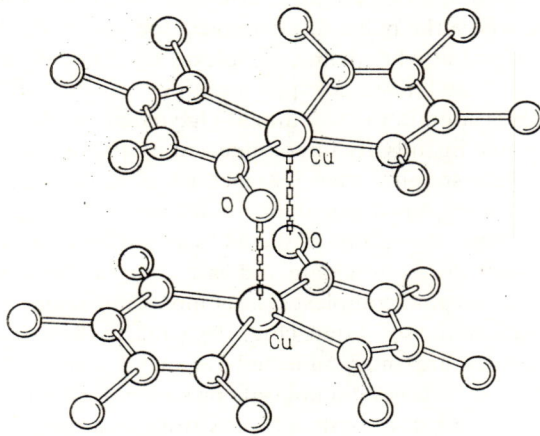

Fig. 11. Structural model of the copper dimethylglyoxime complex. The Cu–O bonds can be seen

Feigl[159] supposed that the strong hydrogen bonds within the palladium complex break and the oxime hydrogens dissociate in alkaline medium. The resulting complex ion with a negative charge of two would obviously dissolve in water. However, the infrared spectrum[98] of palladium dimethylglyoxime showed the presence of quite stable hydrogen bonds. There appears to be no reason why the alkaline medium should abstract protons from these hydrogen bonds, when the same process does not occur with the hydrogen bonds of the corresponding nickel complex of considerably lower stability. Studies by Burger and Dyrssen[89] revealed that the electrically neutral, water-insoluble palladium dimethylglyoxime complex molecule is transformed in alkaline medium to a water-soluble complex with one negative charge. This negative charge originates from the coordination of a hydroxyl ion along the z-axis of the square-planar dimethylglyoxime chelate in alkaline medium. Thus it is clear why the presence of dimethylglyoxime cannot be detected with nickel ions in alkaline solution of palladium dimethylglyoxime: the increase of pH results only in the coordination of a hydroxide ion and does not release dimethylglyoxime.

Alimarin *et al.* have observed an analogous phenomenon in the complex systems of palladium with various dioximes.[10, 58, 429]

Recent examinations revealed why the nickel complexes of dioximes are insoluble in water and why the analogous dioxime complexes of the other transition metal ions are relatively freely soluble in water. Banks *et al.*[24, 25, 27] explained the strikingly low solubility of the nickel complex by the nickel–nickel bonds in the solid precipitate. Studies by Burger and Ruff[86, 95, 97] and Burger and Pintér[93] showed that with the exception of the nickel chelate the transition metal dimethylglyoxime complexes are capable of binding solvent molecules or monodentate ligands along the z-axis. Such monodentate ligands are the hydroxide, halide and pseudo-halide ions, amines, etc. The complexes which more readily form mixed complexes, including the coordination of water molecules, are evidently considerably more water-soluble than the analogous nickel complexes which cannot bind either solvent molecules or monodentate ligands.

Some of the above-mentioned mixed complexes also find application in selective analytical procedures. Thus, the red colour of the mixed complex of iron(II) dimethylglyoxime and ammonia[233] made possible the selective spectrophotometric determination of iron(II). Similarly, the mixed complex of cobalt dimethylglyoxime and iodide was utilized for the micro-determination of cobalt(II).[96]

In connection with the analytical selectivity of dimethylglyoxime, its nickel(IV) complex is of interest. In the presence of excess oxidizing agent the reaction of nickel ions and dimethylglyoxime gives a red complex which is soluble in water.[156, 381] The sensitivity of this reaction is fifty times as great as that of the formation of the nickel(II) dimethylglyoxime precipitate. There are as yet no exact data concerning the stability of this complex, but it may be informative that the nickel(IV) complex is formed by oxidation in the presence of cyanide ions of the high stability tetracyanonickelate(II) complex.

In order to avoid the experimental difficulties resulting from the relatively low solubility of dimethylglyoxime in water, a number of water-soluble dioxime reagents have been synthesized. Of these, cyclohexanedione dioxime (Nioxime) and recently cycloheptanedione dioxime (Heptoxime) have proved to be particularly suitable.[341]

α-Hydroxyaldoximes

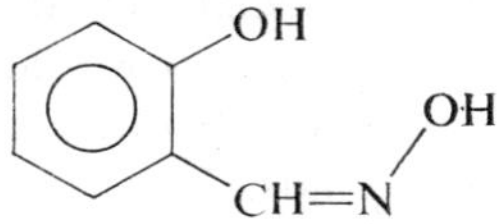

As early as in 1930 Ephraim[147] showed that with copper(II) salicylaldoxime gives a pale yellow precipitate, insoluble even in strong mineral acidic media. Further investigations by Ephraim revealed that all those ligands which contain the functional groups characteristic of salicylaldoxime, react similarly with copper(II), e.g. 2-naphthol-1-aldoxime, 1-naphthol-2-aldoxime, vanilline oxime, quinacetophenonoxime, etc. However, these reagents are not superior to salicylaldoxime in any analytical respect. The analytical selectivity of the ligand arises in connection with the outstanding stability of its complexes with copper(II) and palladium(II) (cf. Table 14). Although the reagent reacts with all $3d^5$–$3d^{10}$ transition metal ions to give complexes which are insoluble in water, only the copper

TABLE 14. LOGARITHMS OF THE COMPLEX PRODUCTS OF SOME SALICYLALDOXIME CHELATES (25°C; in 75% DIOXAN)[90]

Central atom	$\log \beta_2$
Mn(II)	11.9
Fe(II)	16.7
Co(II)	13.5
Ni(II)	14.3
Cu(II)	21.5
Zn(II)	13.5

and palladium complexes form in the presence of acetic acid or in dilute mineral acidic media. (In other words, only these chelates are sufficiently stable to form at such low pH values, i.e. with such low concentration of the free ligand.) Table 14 shows that the stability of the copper complex is five orders of magnitude higher than that of the next most stable complex, the iron(II) chelate. The stability of the palladium(II) complex is not exactly known as yet, but qualitative experiments indicate that it is certainly higher even than that of the copper complex.

Another functional group quite similar to that of salicylaldoxime (which as we have seen above is specific for copper) is α-*acyloinoxime*:

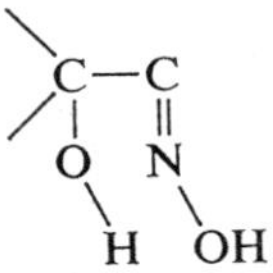

Ligands containing this group react with copper(II) to give precipitates which are insoluble in water and are usually green.[161] The analytical applicability of the reagent is almost independent of the molecular background, i.e. those parts of the molecule to which the functional groups are attached, e.g. phenyl, furoyl, benzyl. Attachment of methoxy, isopropyl or methylenedioxy groups to the benzene ring of benzoinoxime does not affect the analytical selectivity. Hence, in this case the low solubility and the colour of the copper complex, and also the possibility of complex formation with other ions, are determined by the chelate rings and not by the other substituents of the ligand.

The copper(II) complexes of acyloinoximes differ from the salicylaldoxime complexes as regards composition; while the metal:ligand ratio is 1:1 in the former, it is always 1:2 in the latter. In salicylaldoxime complexes the two ligands are interconnected not only through the central atom but by two hydrogen bonds as well (Fig. 12). The hydrogen bonds ensure the higher stability of the MA_2 complex. This is indicated by the unusual order of the stepwise stability constants of the corresponding complexes: $K_2 > K_1$ (cf. Table 8).

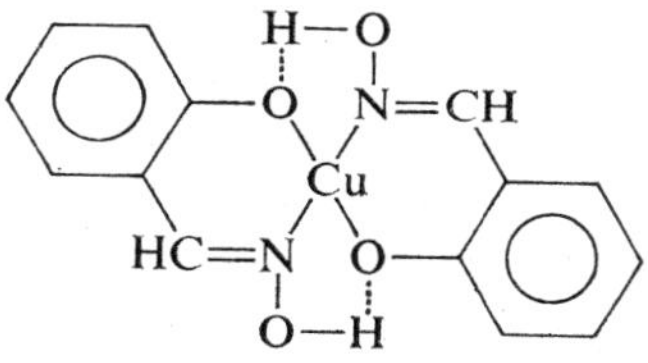

FIG. 12. Structural formula of the copper(II) salicylaldoxime complex

From an analytical point of view a great difference between the two types of copper complex is that both the exact composition and the structure of the salicylaldoxime complex are known and thus it may be utilized in gravimetric determinations. In contrast, the structure and composition of the α-acyloinoxime complexes are not exactly known. Hence their application in quantitative analytical procedures is accompanied by some degree of uncertainty. They are usually applied for the detection of traces of copper.

It should be mentioned at this point that in special circumstances benzoinoxime gives precipitate with cobalt, nickel, palladium and platinum. However, in these complexes benzoinoxime acts as a monodentate ligand,[232] in contrast with the case of the copper complex. According to the studies of Knowles[251] the reagent gives heteropoly compounds of unknown composition with molybdate, tungstate, vanadate and uranyl salts in mineral acidic solutions.

Pyrogallol

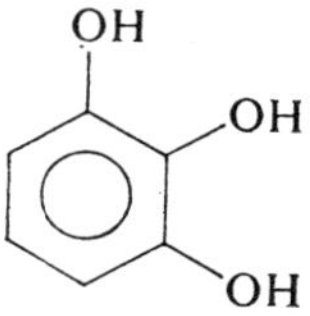

Feigl[155, 160] was the first to point out the analytical significance of pyrogallol. His investigations showed that this ligand precipitates antimony(III) and

bismuth(III) quantitatively from strong mineral acidic media. In acidic solutions pyrogallol does not give a precipitate with any other metal ion and hence it is selective for antimony and bismuth. The precipitates can be utilized for the gravimetric determination of these metals.

Feigl also pointed out that the selectivity of pyrogallol is connected with the three hydroxyl groups; for instance, pyrocatechol and its isomers do not give a water-insoluble precipitate with antimony. The reason for the phenomenon is that these precipitates are polynuclear complexes.[339] The third hydroxyl group is responsible for the interconnection of two complex units.

Pyrogallol also possesses importance as a spectrophotometric reagent in the determination of niobium and tantalum. The absorptions of niobium and tantalum complexes are usually measured at 340 and 335 nm, respectively. The niobium complex is formed in slightly acidic medium, and the tantalum complex in strongly acidic medium (4 N HCl). The absorption spectra are pH-dependent.[339]

β- and o-Amino acids

In β- and o-amino acids the relative positions of the amino and carboxylic groups are such that a metal ion attached to both the amino nitrogen and the carboxylate oxygen atoms closes a six-membered chelate ring. Among complexating agents of this type anthranilic acid has proved to be the most useful in analytical practice. This ligand gives chelate complexes with a variety of metal ions and therefore its analytical selectivity is limited. The analytical selectivity of aliphatic amino acids appears somewhat higher. On the other hand, the stabilities of these complexes, particularly in strongly acidic media, are lower. Due to its outstanding stability, the formation of copper anthranilate possesses a certain selectivity (Table 15); the majority of metal anthranilates form at pH 4.7–5.2, while the formation of the considerably more stable copper complex is already quantitative at pH 2.1.

TABLE 15. NEGATIVE LOGARITHMS OF THE SOLUBILITIES OF SOME WATER-INSOLUBLE ANTHRANILATE COMPLEXES[275]

Central atom	$-\log K_{so}$
Co(II)	10.97
Ni(II)	11.72
Cu(II)	14.18
Zn(II)	9.75
Cd(II)	8.39
Pb(II)	9.81

The analytical significance of this reagent is further increased by the fact that the anthranilic acid content of the complexes of stoichiometric composition may be measured oxidimetrically.[152, 182, 204] Due to the low equivalent weight of the metal ion in such reactions, the oxidimetric methods of this type have proved advantageous for the determination of micro amounts of metals.

α-Hydroxycarboxylic acids

$$-\overset{|}{\underset{OH}{C}}-\overset{|}{\underset{OH}{C}}=O$$

Among α-hydroxycarboxylic acids only mandelic acid possesses considerable analytical importance. Kumins[259] observed that mandelic acid (phenylglycollic acid) precipitates zirconium quantitatively from aqueous media in the form of the tetramandelato complex. The complex is so stable that mandelic acid abstracts zirconium from even its relatively stable sulfate complex. It is known that zirconium can be masked with sulfate in many of its characteristic reactions.

The formation of the mandelo complex of zirconium is an almost fully specific reaction. None of the following metals interfere with the determination of zirconium with mandelic acid: iron, aluminium, chromium, titanium, tin, bismuth, thorium, tellurium, copper, cadmium, vanadium and alkaline earth metals.

As regards the structure of the zirconium complex it is known that two of the mandelic acid moieties act as bidentate, and the other two as monodentate ligands.

Schneer and Hartmann have successfully applied the permanganometric[408] and chromatometric[409] oxidation of the mandelic acid content of the complex precipitate for the determination of micro amounts of zirconium, providing further evidence for the stoichiometric composition of the precipitate. Their method has proved suitable for the determination of zirconium in red mud, indicating the selectivity of the reaction.

Rhodanine and its derivatives

$$S=C\underset{S}{\overset{HN-C=O}{\diagup\diagdown}}CH_2$$

In the case of reagents containing several potential donor atoms, one of the first steps in the study of complex formation is the specification of the functional groups which actually participate in the complexation reaction. The literature on rhodanine derivatives provides a good example of such examinations.

Various organic compounds containing the rhodanine ring give a pale yellow precipitate with silver in dilute nitric acid media. By systematic blocking of the various functional groups in the rhodanine ring it could be decided that the reactivity towards silver is associated with the neighbouring CS and NH groups of the ring. Replacement of the ring sulfur atom for oxygen, or similarly of the CO and CH_2 groups, left the analytical behaviour of the reagent intact. What is more, by introduction of substituents which do not affect the selectivity, the sensitivity of the reagent could be enhanced. In this way Feigl[157] came upon the dimethylaminobenzaldehyde derivative of rhodanine, *p-dimethylaminobenzylidenerhodanine*, which possesses optimum features from an analytical point of view. This compound reacts in acidic medium with only strongly polarizable metal ions, i.e. with those metals which fall in group *b* in Ahrland's classification. Hence Cu(I), Au(I), Pd(II) and Pt(II) interfere with the detection of silver. Many other heavy metals interfere in neutral or alkaline media.

In addition to the determination of silver, the reagent has proved suitable for measuring palladium(II).[17]

Organic nitro compounds

Alkali metal ions give precipitates with relatively few reagents. Therefore, a particularly valuable feature of some organic nitro compounds is that with potassium, rubidium and cesium they form complexes which are sparingly soluble in water. Such nitro compounds are picric acid, *o*-nitrophenol, 2,4-dinitrophenol, trinitro-*m*-cresol, trinitroresorcinol, 6-chloro-5-nitrotoluene-3-sulfonic acid, etc. In practice dipicrylamine has proved to be the most suitable member of the series:[356]

With potassium it gives a yellow precipitate which is practically insoluble in water. The selectivity is relatively high: the reagent reacts also only with cesium, rubidium, thallium(I) and ammonium ions. In the absence of sodium the reagent is capable of the detection of 3 γ potassium; in the presence of sodium the limit of detection is 25 γ.

Feigl[158] assumed that potassium gives a chelate with the ligand by coordination of one of the nitro oxygen atoms and the nitrogen of the secondary amino group. This assumption has not been confirmed. On the basis of our recent knowledge of chelates this structure does not seem probable.

CHAPTER 2

APPLICATION OF SELECTIVE ORGANIC REAGENTS IN QUANTITATIVE CHEMICAL ANALYSIS

I. THE METHODS OF APPLICATION OF SELECTIVE ORGANIC REAGENTS

Organic reagents which form complexes with metal ions can be utilized for the solution of numerous problems in various fields of analytical chemistry. Thus, they may be applied in

1. gravimetric analysis,
2. volumetric analysis,
3. spectrophotometric and colorimetric analyses,
4. polarography,
5. chromatography.

In the following we shall be concerned primarily with those procedures in which the complexes of a given organic reagent with various metal ions form *the basis of the analytical determination directly*.

However, in dealing with the individual selective analytical procedures, mention will be made of the organic reagents applied as *masking agents* in the given reaction. Comparison of the various analytical procedures will demonstrate clearly that almost any reagent which is used in analysis may act as masking agent in some other procedure. For instance, solutions of polyaminocarboxylic acids, which are the most widely used standards, are also the most powerful masking agents. In this book they will appear in this latter capacity only.

Separation and *enrichment* are important fields of the quantitative analytical application of organic reagents. A certain selectivity of a given reaction already includes the possibility of the application of that reaction for separations. Again, all those complexation reactions, in the course of which the metal to be determined is transferred to another phase (e.g. deposits as a precipitate, or can be extracted with some organic solvent immiscible with water), are applicable for enrichment. The author has tried to discuss the various analytical procedures so as to enable an experienced analyst to apply them not only for the determination of a given metal ion but when necessary for its separation or enrichment also.

1. ORGANIC REAGENTS IN GRAVIMETRIC ANALYSIS

The applicability of a given organic reagent in gravimetric analysis depends primarily upon the solubility of its metal complexes. This is evident even to an inexperienced analyst. The relationship between solubility and solubility product, and the factors affecting solubility are dealt with in textbooks. In general, however, very much less attention is devoted to the dependence of the applicability in gravimetric analysis upon the solubility of the reagent itself. In the optimum case the reagent is highly soluble in water and the complex obtained is practically insoluble. In this case the amount of reagent adsorbed on the complex precipitate will be small and easily removable by washing.

Since some organic reagents are weak acids whose undissociated form is only sparingly soluble in water, their freely-soluble alkali or ammonium salts are advantageously applied in analytical practice. For instance, cupferron is the ammonium salt of nitrosophenylhydroxylamine; the sodium salts of anthranilic acid, quinaldic acid and diethyldithiocarbamic acid have also proved suitable. (In the last case the application of the sodium salt is also made feasible by the lower stability of the undissociated acid compared to that of its sodium salt.) In some cases dimethylglyoxime also is applied in sodium hydroxide solution containing the sodium salt of the reagent. In this way the alcohol necessary for the dissolution of the reagent, and which disadvantageously increases the solubility of the metal complex in the medium, can be dispensed with. Of course, replacement of the organic acid by its alkali salt reduces the error caused by co-precipitation of the excess reagent only if precipitation is carried out under such conditions that the reagent exists in its water-soluble anion form (i.e. at sufficiently high pH). The pH of the solution determines the form in which the reagent exists (i.e. protonated or anion form), irrespective of the form in which it was added to the solution.

Organic reagents which are sparingly soluble in water and which cannot be applied in the form of water-soluble salts either, are usually dissolved in some organic solvent miscible with water. Thus, 8-hydroxyquinoline is usually dissolved in alcohol, acetone, or in some cases in media containing acetic acid: dimethylglyoxime is applied in ethanolic solution. In these cases precipitation takes place in water-organic solvent mixtures. Whenever the proportion of the two solvent components is not suitable, considerable error may arise in the analysis. For instance, if the concentration of water is too high, the medium cannot keep the excess of the reagent in solution, the reagent will contaminate the precipitate and the analysis will have a positive error. If, again, the alcohol concentration is higher than necessary, the medium may dissolve a part of the complex formed, thereby resulting in a negative error of the analytical procedure. Therefore, in such cases the accurate observance of prescriptions either elaborated empirically or calculated from a knowledge of the solubilities of the components in various solvent mixtures is particularly important. Care must be taken in respect of the volumes of both the sample and the reagent solution, the extent of dilution, the composition and amount of the wash solution, etc.

The most important advantage of the application of organic reagents in gravimetric analysis is the high molecular weight of the complex obtained. The solubility in water of electrically neutral complexes containing a relatively large organic moiety is extremely low; this renders possible the quantitative precipitation of even micro amounts of metals from very dilute solutions. As a result of the relatively low metal content of such high molecular weight complexes, the gravimetric calculation factor is very advantageous. Of course, it is essential that the composition of the complex precipitate be stoichiometrically reproducible and the precipitate be brought to constant weight without decomposition. Unfortunately, there are many organic complexes which do not meet these requirements. Thus, for instance, the cobalt(III) complex of α-nitroso-β-naphthol, and the complexes of cupferron are not of exact stoichiometric composition. Under such circumstances the complexes are ignited and weighed as metal oxides, or ignited in a reducing atmosphere and subsequently weighed in the metallic form. Accordingly, the high sensitivity of the organic reagent cannot be exploited with such procedures, as with the formation of micro amounts of precipitate the weight to be weighed will be minute. Therefore the analyst must elaborate analytical procedures in which stoichiometrically reproducible complexes are formed. Thermogravimetry, and in particular its most modern variant, derivatography, permit the determination of the optimum conditions for drying precipitates. An extension of the application of organic reagents in analytical chemistry may be expected from further studies in this field.

Precipitation from homogeneous solution

Organic reagents are of great importance in the field of gravimetric analytical procedures based on precipitation from homogeneous solution too. There are several methods of precipitation from homogeneous solution:

1. The precipitating agent is formed from its components in a solution of the metal ion to be determined. A good example is the precipitation of various oxime chelates, where the precipitating agent is formed by slow interaction of the corresponding ketone or aldehyde and hydroxylamine in a solution of the metal ion, usually nickel[208, 390] or palladium.[240]

2. The organic reagent is formed by decomposition of an appropriate parent compound in a solution of the metal ion to be determined. An example is the liberation of 8-hydroxyquinoline by hydrolysis of 8-acetoxyquinoline.[391] The rate of the hydrolysis is relatively low, in compliance with the precipitation of the complex from homogeneous solution. By this technique the precipitation from homogeneous solution of the oxine complexes of aluminium,[282] uranium,[73] magnesium,[130] zinc,[234] thorium[455] and indium[235] has been accomplished successfully.

3. The precipitation of metal complexes whose formation requires an alkaline medium may be accomplished by gradual increase of the pH of a solution containing both the metal ion to be determined and the organic reagent. This slow pH change may be achieved by hydrolysis of an appropriate agent, e.g. urea. The precipitate is deposited gradually from the homogeneous solution corresponding to the rate of hydrolysis of the urea.[57]

4. If a metal ion gives relatively less stable complexes with an organic reagent than with ammonia, it may be precipitated from homogeneous solution by gradual removal of the ammonia content (e.g. by boiling) of a concentrated

ammonia solution containing the metal ion to be determined and the organic reagent. In this way, Firsching[162] determined zinc and cadmium in the form of their oxine complexes.

5. In the case of metal chelates soluble in organic solvents miscible with water or in water–organic solvent mixtures, precipitation from homogeneous solution may be accomplished by gradual removal of the organic solvent (e.g. by boiling or evaporation) from a solution of the metal ion and the organic reagent in a mixture of water and the organic solvent. For instance, nickel dimethylglyoxime can be precipitated in this manner from acetone–water solvent mixtures.[236]

A particular advantage of the methods based on this principle is that by means of appropriate buffers the pH of the solution can be adjusted accurately to the desired value. This permits the accomplishment of separations which otherwise could not be performed.

6. In a less frequently applicable procedure the organic reagent is formed *in situ* in a solution of the metal ion to be determined by oxidation of a suitable parent compound. Thus, for instance, tantalum and niobium can be precipitated in the form of their morin complexes by aerial oxidation of 3,3′,4,5,7-pentahydroxyflavanone in solutions of these metal ions.[110]

Enrichment by co-precipitation

Organic reagents are also important in the field of enrichment methods based on co-precipitation.

The enrichment can be performed by precipitation of an organic reagent, containing appropriate functional groups, which is sparingly soluble in water. In such cases the precipitate consists of the metal chelate of stoichiometric composition formed in the reaction and a great excess of the precipitating agent. Thus, for instance, micro amounts of silver can be precipitated with thionalide,[265] uranium with 1-nitroso-2-naphthol,[472] and alkaline earth metals with potassium rhodisonate.[471]

In another enrichment technique, a solution containing the metal ion to be enriched is treated with a metal ion of related chemical behaviour and a suitable organic reagent which precipitates the two metal chelates together. For instance, trace amounts of scandium, yttrium, cerium, indium and iron can be precipitated together with aluminium oxinate. In such cases it is advisable to perform the precipitation from homogeneous solution, e.g. by slow evaporation of acetone-containing aluminium oxinate solution.[23, 276]

Alimarin and Gibalo[5] have pointed out that co-precipitation of this type frequently gives rise to the formation of polynuclear complexes of stoichiometric composition. For instance: co-precipitation with lumogallione of scandium, yttrium or lanthanum and iron(III) results in a binuclear complex of composition $Fe:M:L$, where $M = Sc$, Y or La and $L = $ lumogallione.

2. ORGANIC REAGENTS IN VOLUMETRIC ANALYSIS

In volumetric analysis organic reagents may be used in either of two basic procedures:

1. The organic reagent is the solute of a standard solution. Complexometric titrations belong here. In these procedures complex-forming reagents are used not only as the solute of the standard solution (e.g. EDTA), but also as metal indicators (e.g. murexide, eriochrome black T, etc.) and in many cases as organic masking agents to retain in solution otherwise hydrolysable salts (e.g. tartrate, citrate, etc.). The theoretical principles of these volumetric methods are dealt with in many monographs[358, 389, 415] and for this reason and because of the limited length of the present book we shall not deal with them here.

However, mention should be made at this point of a new method of indication of the equivalence point, which does not appear in monographs and reviews published so far. It is based on the application of a complex-forming stereospecific organic compound as the solute of the standard solution.[248] In essence, the method consists in titration of the metal ion to be determined with some optically active organic ligand and measurement of the optical rotatory power of the solution. Since both the complex formed and the ligand added with the standard solution are optically active and, moreover, since complex formation affects the optical rotatory power of the ligand, a graph of the optical rotatory power of the solution versus the volume of standard solution added gives a curve which is useful for the evaluation of the equivalence point of the titration. The procedure can be utilized in the concentration range 0.1 to 0.001 M with satisfactory results.

Kirschner,[248] who introduced the procedure, used L-hystidine hydrochloride as stereospecific ligand and examined the formation of the complex with copper. Palma et al.[332] applied the method for the titration of transition metal ions with D-(-)-1,2-propylenediaminetetraacetic acid. The reagent forms complexes with transition metal ions similarly to EDTA. Recording of the titration curve was achieved by means of a photoelectric polarimeter.

Likewise, Reinbold[375] used D-(')-trans-1,2-diaminocyclohexane-N,N',N'-tetraacetic acid for the titration of lead. In determinations of 2–20 mg lead the relative error of the method did not exceed ±0.5%.

2. The organic moiety of the precipitate, which is equivalent to the metal content of the complex precipitated quantitatively with the organic reagent, can be determined titrimetrically. For instance, the organic ligand may be measured by total oxidation, e.g. cinnamic acid is oxidized by sulfuric acid bichromate standard solution to carbon dioxide and water,[409] or some functional group of the ligand can be oxidized oxidimetrically. An example of the latter principle is the determination of hydroxylamine, formed in the acidic hydrolysis of dimethylglyoxime, with bromine chloride standard solution.[85]

In titrations of this type the equivalent weight of the metal ion is usually very small. For instance, in the above-mentioned bromochlorometric measurement of dimethylglyoxime when applied for the determination of nickel, the equivalent weight is equal to the one twenty-fourth of the atomic weight of nickel. (One molecule of the nickel complex contains two molecules of dimethylglyoxime and in the course of the hydrolysis of one dimethylglyoxime two molecules of hydroxylamine are liberated. The bromochlorometric oxidation of the nitrogen atom of hydroxylamine with oxidation number −I gives nitric acid, the process being accompanied by a change of six electrons. A nickel atom is equivalent to four hydroxylamine molecules. Thus the equivalent weight is one twenty-fourth of the atomic weight of the nickel atom.) In the case of determinations based on total oxidation the equivalent weight may be even smaller.[409] By application of dilute (0.01 N) standard solutions even γ amounts of metal ions can be measured with consumptions of several milliliters (i.e. semimicro dimensions).

Such procedures belong among amplification methods.[42] By introduction of new organic reagents and by elaboration of new procedures for the measurement of well-established reagents the field of application of these really microanalytical procedures can be further extended.

However, in connection with these types of determinations, the analyst must keep in mind that the sensitivity of the procedure (i.e. the lower limit of metal concentration which can be determined accurately) depends mainly upon the solubility of the complex precipitate and not on the volumetric procedure elaborated for the measurement of the organic ligand. If the precipitation of the complex is not quantitative, no accurate, reliable analysis data will be available even with perfect methods for the measurement of the ligand in the precipitate. The amplification analytical procedures also amplify the error, originating from the solubility of the precipitate.

Fortunately, the solubility of many metal complexes formed with various organic reagents is so low that the chance offered by the outlined titrimetric methods can be exploited in practice. In addition to the two examples mentioned above, the bromatometric measurement of the 8-hydroxyquinoline content of oxine chelates[47] and of the anthranilic acid content of anthranilate complexes[134] has proved to be suitable.

3. ORGANIC REAGENTS IN SPECTROPHOTOMETRIC ANALYSIS

In many cases metal complexes of organic reagents are coloured. Whenever the coloured complex can be dissolved and the light absorption of the solution follows the Beer–Lambert Law, the colour can be utilized for the spectrophotometric or colorimetric measurement of the concentration. Organic reagents used in spectrophotometry may be classified into two main groups:

1. Ligands giving *water-soluble* complexes with metal ions.
2. Ligands giving metal complexes which are *soluble in apolar organic solvents.*

The most important factors which determine the water-solubility of complexes have been reviewed in an earlier section (pp. 49–51). The solubility in water of the analytically utilized complexes of organic ligands, which are usually of relatively large molecular weight, depends primarily upon the charge of the complex. Cationic and anionic complexes are soluble in water. In contrast with a few exceptions, electrically neutral complex molecules are practically insoluble in water, but readily dissolve in organic solvents.

1. In terms of the above classification, the first group of ligands includes the chargeless polyamines (e.g. ethylenediamine, dipyridyl, terpyridyl, *o*-phenanthroline, etc.). With metal ions these give positively charged water-soluble complex cations. Whenever a complex absorbs light in any region of the spectrum where the ligand does not, the ligand is useful as a spectrophotometric reagent. Taking

into consideration those characteristics of the electron excitation spectra of complexes which were surveyed in an earlier section (pp. 51–53), it is evident that the most useful spectrophotometric reagents are those ligands which give complexes whose spectra contain metal → ligand or ligand → metal charge-transfer bands. These absorption bands are usually well separated from the absorption bands of the ligand itself (i.e. the bands associated with various electron transfers within the ligand molecule). The intensities (extinction coefficients) of these bands are generally higher than those of d–d transition bands which also lie separate from the ligand bands. Hence, in spectrophotometric measurements carried out in the spectral region of the charge-transfer band of a complex, the excess ligand which must be applied in order to ensure the quantitative formation of the complex does not interfere with the analytical procedure. The relatively high intensity of the band permits the elaboration of sensitive analytical procedures.

With a knowledge of the extinction coefficient of the complex at the wavelength of the measurement, the concentration of the solution or the amount of the metal ion in the solution are obtained from the measured extinction value simply by multiplication. In analytical laboratories it is not always possible to measure the extinction exactly at the wavelength of the extinction maximum. However, this does not necessarily cause considerable error if the measurement is carried out in the vicinity of the maximum, or at any rate at a wavelength value where the increase of the absorption curve is not too steep. The value of the extinction coefficient obviously varies with the wavelength at which the measurement is carried out and it also depends greatly upon the monochromaticity of the light applied (the slit width). Therefore, instead of taking the extinction coefficient values given in the literature (there are many contradictory data because of the reasons mentioned), it is advisable to determine accurately the actual multiplication factor to be used for the calculation of the concentration for the given instrument and system by application of a series of solutions of exactly known concentrations. Because of similar considerations, the present book does not in general give multiplication factors in the prescriptions of photometric methods. (For instance, the value of an extinction coefficient determined with a prismatic or grating spectrophotometer is not applicable to another instrument which includes filters, and vice versa.) Though the extinction coefficients which are available in the literature are given in the corresponding tables in this book, as a result of the above considerations it is advisable to use these as informatory data only.

2. In the case of those organic reagents which fall in the second group in the above classification, the ligand is usually formed by dissociation of a proton from the organic reagent and is accordingly a negatively charged ion. Of these many coordinate to the metal ion to compensate its positive charge. In the case of divalent and trivalent metal ions, for instance, complexes MA_2 and MA_3 are formed, respectively. It should be mentioned that ligand ions with one negative charge which belong to this group usually contain two or possibly three donor atoms in an arrangement which permits the formation of chelates. Thus, the coordination of ligands is accompanied not only by compensation of the electric charge of the central ion, but also by the saturation of its coordination sphere. Such an organic reagent, for example, is 8-hydroxyquinoline, where the actual ligand is the 8-hydroxyquinolate ion bearing one negative charge and possessing an oxygen and a nitrogen donor atom. In the 8-hydroxyquinoline complex of Al(III) three ligands

are linked to the central ion, thus compensating the charge of the Al(III) and filling its six coordination sites.

Electrically neutral and coordinatively saturated metal complexes are sparingly soluble in water and quite soluble in organic solvents. Hence metal ions can be extracted in the form of such complexes from their aqueous solutions by organic solvents immiscible with water.[312]

In the case when the chelate-forming ligand neutralizes the positive charge of the metal ion during formation of the complex, but does not saturate the coordination sphere of the central atom, electrically neutral, uncharged donor molecules facilitate the extraction of the coordinatively unsaturated chelate. The phenomenon based on the formation of mixed complexes is termed the *synergism* or the synergetic effect of the two ligands.[61] Under analogous conditions negatively charged hydrophilic ligands hinder the extraction of the coordinatively unsaturated complex, and are said to exhibit an antagonist effect.

If a complex in an organic solvent absorbs in any region of the spectrum where the excess ligand transferred to the organic phase does not, or when the free ligand is not transferred to the organic phase, and if the absorption of the complex follows the Beer–Lambert Law, the reagent may be used for the spectrophotometric determination of the metal ion.

With spectrophotometric methods either based on the absorption of water-soluble complexes or combined with solvent extraction, considerable errors may arise if more than one of the successively formed complex species are present in the solution in significant amounts. Thus, for instance, if in an analytical procedure the absorption of a complex MA is measured at the wavelength of an absorption band characteristic of this species, care must be taken that the excess of the ligand should not be sufficient for the complex MA_2 to form in considerable amount also. The greater the difference between the absorptions of the complexes MA and MA_2 at the wavelength, the greater the error caused by the presence of the complex MA_2 in a given concentration. Nor may we expect the spectra of successively formed complex species to be identical even within narrow spectral regions.

An analogous source of error must also be taken into account when the absorption of the complex MA_2 is to be utilized in the analysis. In this case the analyst should be careful to provide the necessary excess of the ligand so as to minimize the concentration of the MA complex; however, it should not be high enough to permit the formation of the complex MA_3 in considerable amount.

The presence together of successively formed complexes in the solution may cause particularly serious errors in spectrophotometric measurements combined with solvent extraction. Only one of the successive complexes, that which is electrically neutral, can be extracted by the organic solvent. Those complex species which contain less or more ligands than the number necessary to compensate the positive charge of the central metal ion and are accordingly positively or negatively charged, respectively, will remain in the aqueous phase. The possibility of this potential error is least with those complexes in which the compensation of the electric charge of the central metal ion and the saturation of its coordination sphere require the coordination of the same number of ligands.

These considerations clearly demonstrate that those complex systems are the most advantageous for analytical purposes in which for some reason one of the successively formed species is particularly stable. (It should be added that this refers not only to spectrophotometric analytical procedures, but more or less to all types of analysis.) In such complex systems only one complex species dominates within a relatively broad ligand concentration range and this may serve as the basis for the analytical procedure.

4. ORGANIC REAGENTS IN POLAROGRAPHY

It is known that complex formation is accompanied by a shift of the polarographic wave of metal ions towards more negative potentials. Thus the effect of organic reagents is to change the values of the deposition potentials (half-wave potentials) of individual metal ions. With particularly stable complexes it may even be that the polarographic wave of the metal ion does not appear at all within the potential interval available. Hence organic reagents may be useful for the masking of certain metal ions. If a metal ion reducible at a relatively positive potential gives with a certain ligand a more stable complex than another metal ion, which is, in turn, reducible at a more negative potential, the order of the polarographic waves of the two metal ions (i.e. their complexes) may be reversed. In this way the effect of the metal ion, originally depositing at the more positive potential value, interfering with the determination of the other metal can be eliminated (Fig. 13). For example, the polarographic determination of cobalt(II) is generally interfered with by the presence of nickel(II). Nickel is reduced in almost all usual supporting electrolyte solutions at a more positive potential than cobalt (Table 16). However, due to the higher stability of the citrate complex of nickel, the latter does not reduce polarographically in alkaline solution containing citrate ions.[294] Thus cobalt can be determined from the cobalt wave appearing at -1.145 V even in the presence of relatively large amounts of nickel. Similarly, zinc can be determined in the presence of cobalt in alkaline solutions containing EDTA as masking agent. The half-wave potentials of these two metals are almost identical. However, as a result of the high stability of the EDTA complex of cobalt, this ion does not give any polarographic wave in alkaline solution in the presence of EDTA. Under similar conditions, zinc is reduced and its wave

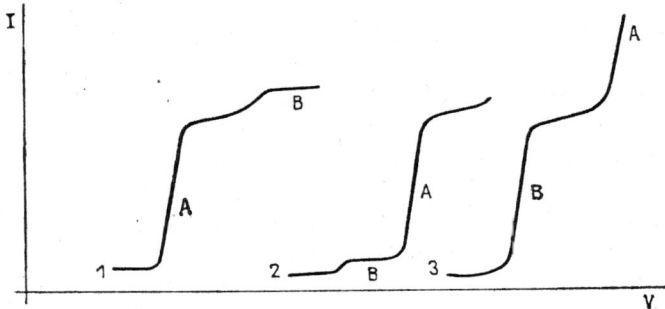

FIG. 13. Effect of complex formation on the polarogram of the metal ion: (1) polarogram of metal ions A and B (the B wave is practically impossible to evaluate in the presence of a large excess of A); (2) waves A and B are interchanged in the presence of complexing agent; (3) recording the polarogram at higher polarograph sensitivity makes wave B evaluable

TABLE 16. HALF-WAVE POTENTIAL, V, OF SOME BIVALENT TRANSITION METALS IN DIFFERENT SUPPORTING ELECTROLYTE SOLUTIONS

Supporting electrolyte solution		Co(II)	Ni(II)	Zn(II)	Mn(II)
$CH_3COOH-CH_3COONa$	(2 M)	−1.19	−1.1	−1.1	—
EDTA (0.1 M) −CH_3COONa	(2 M)	—	—	−1.4	—
KCl	(0.1 M)	−1.20	−1.1	−1.0	−1.55
NH_3-NH_4Cl	(1 M)	−1.29	−0.92	−1.35	−1.66
Sodium citrate (1 M) −NaOH	(0.1 M)	−1.45	—	−1.43	—

appearing at −1.4 V half-wave potential is useful for the determination of zinc in the presence of cobalt.[209]

The great analytical significance of polarographic supporting electrolyte solutions containing organic reagents is clearly indicated by the problem of determining manganese(II). Because of the very negative half-wave potential of manganese(II) (Table 16) almost all metal ions interfere with its determination. Verdier[466] pointed out that with 2 M NaOH solution containing 0.25 M sodium tartrate and Mn(II) an *anodic* wave appears at −0.39 V, which corresponds to the oxidation process Mn(II) → Mn(III). This anodic wave can be utilized for the determination of manganese(II) even in the presence of relatively large amounts of iron, nickel and cobalt. The tartrate complexing agent ensures the stabilization of the manganese(III) in the solution. A similar result was attained by Novak et al.[327] with triethanolamine in the presence of which the half-wave potential of the anodic wave of manganese(II) appears at −0.45 V.

The application of organic reagents in polarography is also useful in the case when a metal complex possesses some electrochemical feature which is different from that of the metal ion and the free ligand. For instance, the dimethylglyoxime complex of cobalt(II) catalyses the deposition of hydrogen at the dropping mercury electrode in unbuffered solutions at pH 6.[449] Cobalt(II) alone does not possess this property and though the positively charged protonated form of dimethylglyoxime catalyses the deposition of hydrogen, in solutions of pH 6 the concentration of protonated dimethylglyoxime which acts as catalyst is so low that it recombines only in the presence of buffer. Thus in unbuffered solutions of dimethylglyoxime the catalytic hydrogen wave does not appear.[100] Hence the catalytic hydrogen wave of cobalt(II) dimethylglyoxime complex may be used for the quantitative determination of cobalt. Of course, the accuracy of the determination depends on the reproducibility of the catalytic wave. It is known that both the shapes and magnitudes of catalytic waves are greatly dependent upon the composition of the solution (ionic strength, the nature of the buffer, pH), the temperature, etc. The reproducibility is not better than ±10% even under optimum conditions. However, because of its specificity, the procedure is of practical importance despite this relatively serious inaccuracy. γ amounts of cobalt can be measured by means of the catalytic hydrogen wave in the presence of hundred-fold excesses of nickel and copper and even larger amounts of other metal ions.[100] A certain selectivity is provided to the procedure by the selectivity with which dimethylglyoxime reacts with various metal ions, and this is increased by the fact that some of the dimethylglyoxime complexes separate from aqueous medium as precipitates and thus do not interfere with the analytical procedure. The procedure is specific since the cobalt(II) complex is the only one of the water-soluble transition metal dimethylglyoxime complexes that catalyses the deposition of hydrogen.

The third manner of the polarographic application of organic reagents is the polarographic determination of the organic component obtained in amounts equivalent to the metal content by precipitation of the metal ion with excess organic reagent and dissolution of the complex precipitate in some appropriate way. For instance, dimethylglyoxime gives an eight-electron reduction wave in hydrochloric acid solution (pH < 4) (it is reduced to 2,3-diaminobutane).[439] Subsequent to precipitation in the usual manner and washing, the nickel(II) dimethylglyoxime precipitate can be dissolved in hydrochloric acid and the dimethylglyoxime content of the solution, which is equivalent to the nickel content to be measured, determined polarographically in the presence of a sodium perchlorate carrier electrolyte.[100] As two ligands are attached to the nickel(II) in the complex, a nickel atom results in a sixteen-electron reduction wave. Hence in this respect the method is eight times as sensitive as the direct polarographic determination of nickel. Actually, the sensitivity of the procedure is limited not so much by the sensitivity of the polarographic measurement but by the solubility of the precipitate. The main advantage of the method lies not in its greater sensitivity, but primarily in the selectivity of the formation of the nickel(II) dimethylglyoxime complex.

In some cases the application of organic reagents renders possible the determination of metals whose direct polarographic determination encounters serious difficulties. For example, the direct polarographic determination of aluminium(III) is well known to be troublesome; an aluminium wave utilizable for quantitative purposes can be gained only within a quite narrow pH range (pH 3.2–3.5).[365] In more acidic solutions the hydrogen wave overlaps the aluminium wave, while in more alkaline medium the hydrolysis of the metal ion prevents the determination of aluminium(III). Willard and Dean[478] pointed out that the reduction wave of *Pentachrome-Violet S.W.*, a polarographically reducible *o*-dihydroxy azo dye, appears at a half-wave potential of -0.34 V. This dye gives a complex with aluminium(III) and the dye in the complex reduces at a potential more negative by 0.2 V than in the free state. Accordingly, two polarographic waves appear in the case of solutions containing an excess of the dye (in a 0.2 N acetic acid–sodium acetate buffer of pH 4.6). The heights of the first and second waves are proportional to the concentrations of the free dye and the complexed aluminium(III), respectively. By this method aluminium can be determined within the concentration range 0.01–0.3 mg Al/50 ml. Perkins and Reynolds[338] have similarly achieved the polarographic determination of aluminium with the dye *Solochrome Violet R. S.*

These and related determinations with azo dye ligands possess the further advantage that complex formation is not accompanied by deposition of precipitate and hence the adsorption of the polarographically active ligand upon the precipitated complex does not appear among the potential sources of error. Serious errors may arise from adsorption in those cases where the decrease in concentration of a polarographically reducible reagent is measured in the course of the formation of one of its metal complexes when the latter is precipitated. Among polarographically reducible organic reagents are, for instance, 8-hydroxyquinoline and picrolonic acid. These have been utilized for the polarographic determination of magnesium(II) and calcium(II), respectively.[294] The metal ions were precipitated by the organic reagent the excess of which was measured polarographically. However, due to the adsorption of the ligand on the precipitate, these procedures possess a relatively high positive error.

As a quite recent field of the polarographic application of organic reagents, mention must be made of the examinations of Brainina et al.[74] By either reduction or oxidation on a graphite electrode, they transformed the metal ion to be determined into an appropriate oxidation state so as to give a highly stable complex with the ligand present; the complex deposited on the electrode surface in the form of an insoluble precipitate. The determination was carried out by electrolytic dissolution of the precipitate thus collected on the electrode surface. In the course of the elaboration of the procedure the applicability of 1-nitroso-2-naphthol, Rhodanine C and ethylenediamine were examined. The main advantage of the procedure is that the determination is accompanied by enrichment of the component to be determined, and this renders possible the elaboration of micro methods.

5. ORGANIC REAGENTS IN CHROMATOGRAPHY AND ION EXCHANGE

Complex formation may also fundamentally alter the chromatographic and ion exchange behaviour of metal ions. Such changes occur primarily in those cases when complex formation is accompanied by a change in the electric charge of the metal ion. By means of appropriate complexing agents (e.g. EDTA) metal cations can be transformed to complex anions. These can no longer be bound by cation-exchange resins, but only by anion-exchange resins. Complexation may also result in the formation of chargeless species. With suitable solvents these may pass both types of ion-exchange resins without being bonded.

Hence the efficiency of chromatographic and ion-exchange separation methods can be enhanced by organic ligands. The role of the complex-forming organic reagent may be various:

1. Elution with a solution of some complex-forming reagent. In this case the unequal stabilities of the complexes formed may make possible the accomplishment of separations which could not be effected merely on the basis of the selectivity of the chromatographic adsorbent or the ion-exchange resin itself. For instance, Blaedel and Olsen[60] separated the components of a radioactive decomposition product containing 36 different radioactive elements by chromatography using solutions of various organic complexing reagents for the elution. With a suitable complexing agent of weakly acidic character and by proper choice of the pH of the solution certain metal ions will form complexes with the organic reagent while others will not. The greater the stability of the complex of a given metal ion with a ligand molecule dissolved in the eluting solution, the lower the pH limit at which complexation proceeds. By appropriate choice of the organic reagent and variation of the pH of the elution solution even such difficult separations can easily be achieved as the separation of alkaline earth metals[267, 467] or rare earth elements.[11, 122]

2. When a strongly basic anion-exchange resin is saturated with some chelate-forming organic reagent, the resin acts as a cation-exchange resin containing complex-forming active sites and it makes possible the quantitative separation of various metal ions. The reason for this phenomenon is that only some of the functional groups of the polyfunctional complex-forming organic reagent (applied in excess) bind to the anion-exchange resin; the remaining chelate-forming groups are available for the metal ions applied to the column. However, the organic reagent more or less saturates the original functional groups of the anion-exchange resin and thus reduces its anion-exchange property.[76, 392, 394] Thus, for instance, Samuelson[393] applied an anion-exchange resin previously treated with EDTA for the separation of alkali and alkaline earth metals. Above pH 7 alkaline earth metals form stable complexes with EDTA and hence these ions are retained quantitatively by the resin. In contrast, alkali ions do not react with EDTA and hence pass through the column.

3. Organic reagents which give coloured or fluorescent complexes with metal ions are useful for the location of various metals situated at various parts of a chromatographic column or paper chromatogram. For example, in the separation of rare earth metals the various metal ions can be shown up by impregnation of the column with morin.[308]

In connection with the analytical application of ion-exchange resins, the excellent monograph of Inczédy[215, 216] also deals with the role of complexing agents. Instead of further discussion of the topic, we here refer the reader to this work.

II. SOME IMPORTANT ORGANIC METAL REAGENTS

In the course of the solution of various practical analytical problems and the investigation of the general principles of the chemical behaviour of different organic reagents, and in some cases even as "by-products" of other research, analytical chemists have prepared several hundred organic reagents.[460a] For instance, the well-known monograph of Welcher[473] deals with more than 800 reagents and even this source is incomplete as regards the organic compounds applied analytically. However, the majority of these several hundred reagents have not found application in analytical practice. The number of organic reagents successfully applied in analysis by various research teams barely exceeds fifty. Even some of these are utilized only as masking agents or buffer materials. Nevertheless the combined application of these relatively few compounds renders possible the accomplishment of a great many analytical separations and selective determinations of metals.

Whenever a solution contains only one metal ion, it can usually be determined equally satisfactorily and accurately with a number of reagents. In a given analytical determination the practical analyst selects from equally suitable reagents that which is the most useful under the given conditions (accompanying ions, composition of the sample).

Efforts have been made in the compilation of this section to select those reagents and illustrate their use, which have proved the most suitable in the experience of the author. The author has tried to discuss the subject matter so that the possibilities of various separations and selective determinations should be accentuated. However, analytical recipes will be included only for the most important procedures. Most of the analytical procedures given have been applied in the author's laboratory, while others have been taken from publications which the author considers reliable. Some of the original recipes have been modified on the basis of the author's experience. The present reagent classification, like all others, is more or less arbitrary and has been made in part from a consideration of the donor atoms and in part from that of the functional groups.

The author's experience in the field of analytical procedures has led him to the conclusion that spectrophotometric procedures in general, but particularly those combined with solvent extraction, are the most satisfactory for the selective determination of the individual components of composite substances. This is why these methods are dealt with in some detail in the following.

1. PHENANTHROLINE AND RELATED COMPOUNDS

1,10-Phenanthroline

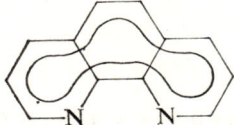

Molecular weight: 180.20
White crystalline material. Melting point 93–94°C.
Solubility: dissolves in 300 parts (by weight) of water or 70 parts of benzene; readily soluble in alcohol and acetone.

1,10-Phenanthroline is the most suitable reagent for the selective determination of iron(II) with which it gives a vivid red complex readily soluble in water.[171] In the region between pH 2 and 9 the intensity of the colour of the complex is independent of the pH of the solution. The high stability of the complex ensures that the colour of the solution remains constant for days and it follows the Beer–Lambert Law within a relatively broad concentration range.

Silver(I) and bismuth(III) give precipitates with the reagent, and some other ions which form complexes with the reagent, e.g. cadmium(II), mercury(II) and zinc(II), may interfere with the determination of iron(II) by consuming the reagent. This interference is eliminated by applying a great excess of the reagent. Thus 0.5–1 γ/ml iron can readily be determined in the presence of a 50-fold amount of cadmium, a 10-fold amount of zinc, but a maximum 5-fold amount of mercury. In the presence of copper the pH of the solution is adjusted to 2.5–5.0, but the concentration of copper may not be more than 5–10 times that of the iron. The presence of nickel alters the absorption spectrum considerably; hence in the presence of this metal in concentrations comparable with that of the iron the obtained data are not reliable. The presence of cobalt(II), tin(II) and tin(IV) in lower concentrations usually does not interfere with the determination provided that the pH of the solution is maintained at 2–3.

The most remarkable advantage of the reagent is that it renders possible the determination of iron in acidic media; thus metal ions which undergo hydrolysis in neutral or slightly alkaline media do not interfere with the procedure. Anions which give highly stable complexes with iron(III), e.g. pyrophosphate, phosphate and fluoride, interfere inasmuch as they hinder the reduction of iron(III) since the determination is based on the colour of the iron(II) complex.

Preparation of the reagent solution. A 0.5% aqueous solution of o-phenanthroline is used. The reagent solution should be stored in the dark and it should be used only while colourless.

Determination of iron. A solution containing 0.01–0.2 mg iron is transferred into a 25 ml volumetric flask, the pH of the solution is adjusted with sodium citrate to 3–4 and 1 ml 1% aqueous hydroquinone solution is added to reduce iron(III). Finally 1 ml 1,10-phenanthroline solution is added to produce the red complex. The absorption of the solution is measured at 508 nm after 1 hour.

The reduction of iron(III) can also be carried out with hydroxylamine. In this case the pH adjustment is done with sodium acetate.[398]

In some cases 1,10-phenanthroline is used as a gravimetric reagent. Nickel, copper, lead and cobalt are precipitated by phenanthroline in the presence of thiocyanate ions in the form of complexes $M(ofen)_2(SCN)_2$. By this method these metals are precipitated quantitatively and selectively from solutions containing alkaline earth metals, scandium, aluminium and chromium.[349, 350, 353, 354]

Neocuproine

(2,9-dimethyl-1,10-phenanthroline)

Molecular weight: 208.25
White crystalline material.
Solubility: sparingly soluble in water, readily soluble in alcohol and acetone.

Due to the steric hindrance of the methyl groups attached to the carbon atoms adjacent to the nitrogen donor atoms of phenanthroline, the reagent does not give the low spin vivid red complex characteristic of phenanthroline derivatives with iron(II). However, in the presence of reducing agents it reacts with copper to give a copper(I) complex of composition MA_2 and of tetrahedral symmetry. This chelate is insoluble in water and can be extracted by chloroform in which the absorption maximum of the complex appears at 457 nm. In this way the concentration of the complex can be measured. The method is suitable for the determination of copper in iron, manganese and vanadium ores even in the presence of aluminium, germanium, titanium and silicon.

Preparation of the reagent solution. Neocuproine is dissolved in abs. ethanol to 0.1% concentration.

Determination of copper. 5 ml 10% hydroxylamine hydrochloride solution and then 10 ml 30% sodium citrate solution are added to 10 ml sample solution containing not more than 200 γ copper. The pH of the solution is adjusted to 4–6 with ammonium hydroxide solution. Subsequent to the addition of 10 ml neocuproine reagent solution the mixture is vigorously shaken with 10 ml chloroform for 30 seconds. After separation of the two layers the aqueous phase is extracted with a further 5 ml chloroform. The absorption maximum of the copper complex appears in chloroform solution at 457 nm.[185]

Neocuproine is today considered one of the most selective reagents. Unfortunately, it is rather expensive. 2,3-bis-(2-pyridyl)quinoxaline, prepared by Belcher *et al.* by condensation of *o*-diketone, 2,2′-dipyridyl and *o*-phenylenediamine[44] contains the functional grouping characteristic of cuproine; it is quite suitable for the determination of copper and it is cheap. Starting with various substituted *o*-phenylene-diamines Belcher[41] synthesized 25 different quinoxaline derivatives, of which 2,3-bis-[2-(-methyl)-pyridyl)]quinoxaline proved to be identical with neocuproine as regards analytical selectivity. Besides copper, titanium(III) is the only metal ion which gives a colour reaction with the reagent. However, the coloured titanium(III) complex is formed only at lower pH and hence it does not interfere with the determination of copper. The copper complex of the reagent can be extracted quantitatively with isopentyl alcohol usually as a perchlorate ion pair.

2,2′-Dipyridyl

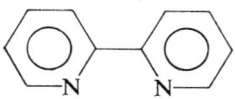

Molecular weight: 156.18
White crystalline material. Melting point 70–71°C.
Solubility: sparingly soluble in water, readily soluble in ethanol, diethyl ether, chloroform and benzene.

With iron(II) the reagent gives a red chelate which is analogous to the corresponding phenanthroline complex and is also useful for the quantitative determination of iron.[314]

Preparation of the reagent solution. Dipyridyl is dissolved in ethanol to 1% final concentration.

Determination of iron(II). 0.40 ml 1% dipyridyl solution and 10 ml 20% aqueous ammonium acetate solution are added to 10 ml sample solution containing 50–100 γ iron and the whole is diluted to 100.0 ml final volume in a volumetric flask. After about 30 minutes standing, the extinction of the solution is measured at 500 nm.
Whenever the overall iron content of a sample is to be measured, the iron(III) content is reduced by addition of 0.02 g ascorbic acid prior to the addition of the dipyridyl reagent solution.

The procedure has been applied for the analysis of many iron-containing pharmaceuticals.[345] In the analysis of not fully colourless products the absorption of the complex is measured against the same amount of the sample itself.

Like *o*-phenanthroline, 2,2′-dipyridyl too may be applied as a gravimetric reagent. Copper, nickel, cobalt and zinc can be precipitated from solutions containing thiocyanate ions in the form of complexes $M(dip)_2(SCN)_2$.[350, 352]

2. OXIMES

Dimethylglyoxime

$$\begin{array}{c} H_3C \\ \diagdown \\ C=N \\ \mid \\ C=N \\ \diagup \\ H_3C \end{array} \quad \begin{array}{c} O-H \\ \diagup \\ \\ \\ \diagdown \\ O-H \end{array}$$

Molecular weight: 116.12
White crystalline powder. Melting point 237–240°C (with decomposition).
Solubility: sparingly soluble in water (0.04 g in 100 ml), somewhat more soluble in ethanol (1 g in 100 ml). Soluble in diethyl ether and acetone.

Under appropriate conditions dimethylglyoxime is specific for nickel(II) and palladium(II), but it also forms coloured water-soluble complexes with iron(II), cobalt(II) and copper(II).[88]

Gravimetric determination of nickel and palladium. Most of the procedures based on the formation of dimethylglyoxime complexes are gravimetric methods.

Preparation of the reagent solution. Usually a 1% solution of dimethylglyoxime in ethanol is applied as reagent solution.

Determination of nickel in the presence of other metal ions. The sample solution containing 20–100 mg nickel in about 150 ml is treated with 25 ml 10% tartaric acid solution. Whenever the solution also contains iron(II), this is oxidized to iron(III) by addition of a few drops of bromine water. Then 1–2 g ammonium chloride is added and the solution heated to 80–85°C. The reagent solution (5 ml for every 10 mg nickel) is poured at this temperature into the sample solution. Care must be taken that the final concentration of alcohol be 30–35% in the reaction mixture. With a lower or higher alcohol concentration in the solution the deposition of the reagent or the solubility of the complex, respectively, adversely affect the result. With continuous stirring of the solution, either with a stream of nitrogen or with a magnetic stirrer, 2 M ammonium hydroxide solution is added in a thin stream until complete precipitation of the complex (about pH 8). After standing for about 1 hour the precipitate is collected on a sintered glass filter, washed with warm water, dried at 130–150°C and weighed. The gravimetric multiplication factor is 0.2032 for nickel.

The optimum drying temperature of the precipitate has been determined by thermogravimetric examinations.[148] The complex is stable, constant in weight and retains its stoichiometric composition up to 200°C, but it decomposes rapidly at 200–260°C.

In the presence of cobalt, manganese and zinc the precipitation is carried out in a sodium acetate–acetic acid buffer solution. In such cases making the solution sufficiently alkaline for the precipitation of the nickel dimethylglyoxime complex is achieved with 20% sodium acetate solution instead of with 2 M ammonium hydroxide.

Determination of palladium. The palladium complex is quantitatively deposited from slightly mineral acidic media (e.g. containing 1% hydrochloric acid). The precipitate is washed with hot water and dried at 110–150°C. According to thermogravimetric examinations the precipitate is stable (constant in weight) up to 240°C. The gravimetric multiplication factor is 0.3167 for palladium.

In measurements of small amounts of palladium the palladium dimethylglyoxime complex is extracted with chloroform and measured spectrophotometrically. The absorption maximum of the complex appears at 380 nm and the absorption follows the Beer–Lambert Law.[132]

Volumetric determination of nickel. Tougarinoff[457] and Furmann and Flagg[184] dissolved the filtered and washed nickel dimethylglyoxime precipitate in acid.

The dimethylglyoxime was hydrolysed by refluxing with the acid to the corresponding ketone and hydroxylamine and the latter was determined oxidimetrically. Tougarinoff oxidized the hydroxylamine with iron(III) to dinitrogen oxide and measured the obtained iron(II) permanganometrically. In this procedure the equivalent weight of nickel is one fourth of the atomic weight. Furmann and Flagg oxidized the hydroxylamine with bromate to nitric acid. In this procedure the equivalent weight of nickel is one twenty-fourth of the atomic weight. In the experience of Furmann and Flagg, as well as of the present author, the procedure is suitable only with solutions of the order of 0.1 N, and with consumptions of less than 10 ml standard solution the error is 4–5%. Satisfactory results are obtained with consumptions over 20 ml (5–10 mg Ni). Hence the advantage offered by the low equivalent weight is lost.[91]

Burger, Gaizer and Schulek elaborated a procedure for the determination of hydroxylamine based on oxidation with bromine chloride. The reaction is described by the equation:

$$NH_2OH + 3\ BrCl + 2\ H_2O = HNO_3 + 3\ HBr + 3\ HCl.$$

The oxidation proved to be suitable for the measurement of hydroxylamine even in 0.01 N concentrations.

Taking into consideration the low solubility of the nickel dimethylglyoxime precipitate, Burger[85] utilized the reaction for the volumetric determination of small amounts of nickel by dissolving the nickel dimethylglyoxime complex precipitate in hydrochloric acid and hydrolysing the dimethylglyoxime to hydroxylamine and the corresponding ketone.

Since each nickel atom reacts with two molecules of dimethylglyoxime in the formation of the precipitate and each dimethylglyoxime molecule liberates two hydroxylamine molecules in the hydrolysis, and since the bromine chloride oxidation of hydroxylamine is accompanied by a change of 6 in the oxidation number of the nitrogen atom ($-I$ to $+V$), the equivalent weight of nickel is one twenty-fourth of the atomic weight in this procedure.

1 ml 0.01 N BrCl solution is equivalent to 24.45 μg nickel.

The selectivity of the procedure is equal to the selectivity of the precipitation of the nickel dimethylglyoxime complex precipitate and hence it makes possible the determination of small amounts of nickel in the presence of almost all other metals.

Preparation of the standard solution. 0.01 N (bromate: 2 bromide) standard solution: 0.2784 g potassium bromate and 0.3967 g potassium bromide are weighed with analytical accuracy and dissolved in distilled water to 1000.0 ml final volume. On acidification the bromate and bromide ions react to give bromine chloride according to the equation:

$$BrO_3^- + 2\ Br^- + 3\ Cl^- + 6\ H^+ = 3\ BrCl + 3\ H_2O.$$

Procedure. The sample solution (max. 5 ml) containing 75–250 γ nickel is placed with analytical accuracy in a 25 ml beaker and 0.2 ml 1% dimethylglyoxime solution in alcohol and 1 ml 2 N ammonium hydroxide solution are added dropwise to the boiling sample. The mixture is gently boiled for 5 minutes, allowed to cool for 10 minutes and poured onto a G3 sintered glass filter. The mother liquor is filtered by suction, the precipitate remaining on the walls of the beaker and in the filter is washed with 2×6 ml of a hot solution of nickel dimethylglyoxime solution (saturated at 20°C) and with 2 ml cold water. Subsequently the precipitate is dissolved in 6 ml hot 20% hydrochloric acid, and the beaker and the filter are washed with 2×6 ml 20% hydrochloric acid. The acidic solutions are collected in a 100 ml beaker, boiled for about 1 minute, allowed to cool for about 10 minutes, washed quantitatively into a *Schulek* bromination flask and diluted with distilled water to 100–150 ml. In order to eliminate the interfering effect of the oxygen dissolved in the reaction mixture, about 0.5 g potassium hydrogen carbonate is added to the solution. (If the complete dissolution of the precipitate requires more hydrochloric acid than specified above, the amount of potassium hydrogen carbonate is increased proportionally.) Then 0.01 N standard solution (bromate: 2 bromide) is pipetted into the reaction mixture so that the bromine chloride formed by the action of the hydrochloric acid is in more than 100% excess. The flask is immediately closed with its moistened adapter furnished with a tap. After standing 15 minutes the flask is cooled with running water under the tap (when the internal pressure falls below 760 mmHg) and 10 ml freshly prepared 5% potassium iodide solution is added to the reaction mixture through the adapter. The iodine liberated is immediately titrated with 0.01 N sodium thiosulfate standard solution in the presence of starch indicator.

1 ml 0.01 N bromine chloride is equivalent to 24.45 γ nickel (log 38,828). The error of the method is -2%.

The maximum 2% negative error appearing in the region of 0.01 N (75–270 γ nickel) originates from the solubility of the precipitate. In the region of 0.1 N (0.75–2.5 mg Ni) the negative error vanishes.

With 0.1 N concentration the procedure is modified as follows:
The precipitation is effected with 2 ml 1% dimethylglyoxime solution in a maximum volume of 6 ml. The pH of the solution is adjusted to the desired value by addition of 2 ml 2 M ammonium hydroxide. The oxidation of hydroxylamine with bromine chloride requires a reaction period of 5 minutes. Otherwise the procedure is unchanged.

I ml 0.1 N BrCl is equivalent to 0.2445 mg nickel (log 38,828). The error of the procedure is $\pm 1\%$.

Solvent extraction determination of nickel. The nickel dimethylglyoxime complex can be extracted by chloroform from the aqueous phase. The distribution ratio of the complex is 320. The absorption maximum appears at 366 nm.

The distribution ratio is defined here and in the following so that the concentration of the complex in the organic phase is in the numerator and that in the aqueous phase in the denominator.

Procedure. 10 ml 50% ammonium citrate solution, ammonium hydroxide (to pH 8) and 20 ml 0.2% dimethylglyoxime in ammonium hydroxide are added to the sample solution (about 100 ml) containing 20–400 γ nickel and the mixture is extracted with 10–15 ml chloroform.[202]

The method is useful for the determination of the nickel contamination of copper. 0.5 g sodium tartrate and 5 g crystalline sodium thiosulfate are dissolved in 10 ml sample solution containing 1 g copper and 5 ml acetic acid–sodium acetate buffer (pH 6.5) is added. Subsequent to the addition of 50 mg hydroxylamine hydrochloride and 2 ml 1% dimethylglyoxime solution in ethanol the mixture is vigorously shaken with 6 ml chloroform for 1 minute to extract the nickel complex. The absorption maximum of the nickel complex in chloroform solution appears at 366 nm.[325]

Spectrophotometric determination of cobalt. The reddish-brown mixed complex (absorption maximum at 435 nm) of cobalt(II) dimethylglyoxime with iodide ions has proved to be suitable for the selective determination of cobalt(II). Under the conditions given the other transition metals do not give similar mixed complexes with iodide ions. The water-soluble dimethylglyoxime parent complexes, apart from the iron(II) dimethylglyoxime complex, do not absorb at 435 nm (the absorption maximum of the mixed complex) even with relatively high concentrations of the complexes. The cobalt(II) mixed complex, which is readily soluble in water, can easily be washed out of the nickel(II) dimethylglyoxime precipitate with water. Within the concentration range $2 \times 10^{-6} - 10^{-3}$ mole/litre the extinction of the complex at 435 nm follows the Beer–Lambert Law exactly; the molar extinction coefficient is $10,600 \pm 100$.

Calcium, magnesium, manganese, zinc and cadmium do not interfere with the determination even in concentrations several hundred times that of the cobalt.

The copper(II) present is reduced by the iodide before addition of the dimethylglyoxime. The resulting copper(I) iodide is dissolved by the excess iodide to give $[CuI_4^{3-}]$. The iodine formed in the reaction is reduced with sodium thiosulfate before addition of the dimethylglyoxime. Thus 0.6–1.2 γ/ml cobalt can be measured with the usual accuracy in the presence of more than 100 γ/ml copper. Copper interferes with the determination only when present in such a high concentration that copper(I) iodide is not dissolved by the excess iodide.

On adjustment of the pH of the solution, aluminium(III) and bismuth(III) precipitate as aluminium hydroxide and bismuth oxyiodide, respectively. The silver content of the solution precipitates as silver iodide. On addition of the dimethylglyoxime the nickel content precipitates as the nickel(II) dimethylglyoxime complex. The cobalt mixed complex, which is readily soluble in water, can easily be washed out of these precipitates. Therefore aluminium, bismuth, silver and nickel ions do not interfere with the determination even in amounts several hundred times that of the cobalt.

The iron(II) dimethylglyoxime complex absorbs strongly at the wavelength of the absorption maximum of the cobalt mixed complex. The interfering effect of iron is eliminated by oxidation of the Fe(II) to Fe(III) and masking of the latter with a great excess of pyrophosphate. In this way even several-thousand-fold amounts of iron do not disturb the procedure.

The coloured iodide complexes of the mercury(II) and lead(II) interfere with the determination of cobalt and hence must be removed from the solution previously. Mercury(II) is removed as mercury sulfide by treatment of the acidified solution with H_2S and lead(II) as lead sulfate by treatment with ammonium sulfate.

Procedure. 0.8 g potassium iodide is dissolved in the sample solution (5–6 ml) containing 2–20 γ cobalt and the pH of the solution is adjusted to 5 with 10% sodium hydrogen carbonate. The fine adjustment of the pH is done by addition of 1.0 ml buffer solution (pH 6). Finally 1 ml 0.1% dimethylglyoxime solution is added and the solution diluted with distilled water to 10.0 ml. The extinction of the solution is measured at 435 nm in a 1 cm cell and the obtained value is multiplied by 5.57 to give the concentration of cobalt in γ/ml units.

In the presence of copper, iodine is liberated on addition of the potassium iodide; it is reduced with a few drops of 0.1 N sodium thiosulfate. Otherwise the above method is followed.

In the presence of aluminium, bismuth, silver and nickel the mixture is filtered through a G3 sintered glass filter before the spectrophotometric measurement. The precipitate is washed with 3×1 ml water.

Whenever the sample to be analysed contains mercury and lead, these are removed as mercury sulfide and lead sulfate, respectively, before the analytical procedure, according to the methods of gravimetric analysis.

In the presence of iron(II) and iron(III) 0.10 g sodium pyrophosphate per 10 ml sample solution is dissolved in the reaction mixture containing the cobalt to be determined and a few drops of bromine water are added to oxidize the iron(II). In a later step the excess of the bromine liberates iodine from the iodide and this is reduced with thiosulfate. Otherwise the standard procedure is followed.

Polarographic determination of cobalt(II) and nickel(II). The cobalt(II) dimethylglyoxime complex exerts a catalytic effect upon the deposition of hydrogen at the dropping mercury electrode.[100]

None of the other dimethylglyoxime complexes gives a wave analogous to the catalytic wave of the cobalt(II) complex. The reason for the phenomenon is presumably the reduction of that hydrogen atom of the complex which participates in the hydrogen bond and accordingly this process is responsible for the catalytic effect of the complex molecule. According to infrared spectroscopic analysis[98] in the divalent transition metal complexes of dimethylglyoxime examined, the hydrogen bond is weakest in the cobalt(II) complex. Under the action of water the catalyst is regenerated and hence a well measurable catalytic current results even in the absence of buffer.

Within that concentration range where the height of the wave is proportional to the concentration of cobalt, the catalytic hydrogen wave can be utilized for the determination of micro amounts (1.5–120 γ/ml) of cobalt. As is well known, both the shape and the height of catalytic waves are extremely sensitive to the composition of the solution.[278] Therefore the cobalt content is determined by the addition method, instead of by means of calibration curves. The polarogram of the sample solution is recorded twice, before and after addition of a known amount of cobalt, and the original cobalt content is calculated from the increase in wave height due to the added cobalt.

Examinations have been carried out concerning the interference by nickel, copper, iron, manganese, mercury, lead, zinc, palladium, calcium and magnesium.[100] Of these, nickel, palladium and mercury separate as insoluble precipitates on preparation of the sample solution (the latter in the form of a basic salt). The precipitate is removed by centrifuging the solution. The procedure is useful even in the presence of 100-fold amounts of these metals. The error of the procedure includes the average error ($\pm 10\%$) of the catalytic wave reproducibility.

The nickel content of the precipitate remaining in the centrifuge tube may be determined by measuring the dimethylglyoxime, which is equivalent to the nickel. In acidic medium dimethylglyoxime is reduced with an eight-electron wave on the dropping mercury electrode.[439] The height of the reduction wave of dimethylglyoxime is proportional to the concentration. The wave height can be evaluated in the presence of nickel also: the nickel(II) dimethylglyoxime precipitate dissolves in acid and at the optimum pH value (1.8) for the measurement of the dimethylglyoxime content, the nickel(II) complex is not formed. By this method 15–180 γ/ml nickel can be determined.

Procedure. A stock solution containing 0.6–10 mg cobalt per 100 ml is prepared from the sample (containing both cobalt and nickel) to be analysed. 10.0 ml of this solution is pipetted into a beaker together with 10 ml 3 M sodium perchlorate solution and 10 ml 0.1 M dimethylglyoxime in ethanol, and the pH of the mixture is adjusted to about 6 with sodium hydroxide. The nickel(II) dimethylglyoxime precipitate is removed by filtration or centrifugation. The precipitate is washed with distilled water until 100 ml filtrate and washings have been collected. The solution is freed from oxygen by passage of a stream of nitrogen gas, and measured polarographically. Another aliquot of the same stock solution is treated with a cobalt(II) solution of known concentration so that the height of the cobalt wave be about twice that of the stock solution, and the polarogram of this solution is also recorded. From the measured increase in wave height and the amount of cobalt(II) added the cobalt content of the stock solution can be calculated.

If the sample to be analysed also contains copper, potassium iodide is dissolved in the mixture before addition of dimethylglyoxime and the iodine liberated is reduced by a few drops of 0.1 M sodium thiosulfate. The dimethylglyoxime is added subsequently, followed by adjustment of the pH. Otherwise the above procedure is followed.

If the nickel content of the sample is also to be determined, the nickel dimethylglyoxime precipitate is separated by centrifugation. The precipitate is dissolved in 0.1 N hydrochloric acid. (The precipitate is washed with the hydrochloric acid solution into a beaker and triturated with a glass rod until completely dissolved.) The homogeneous solution is then diluted with distilled water and hydrochloric acid to about 5–15 γ/ml nickel concentration and pH about 1.5–2. The nickel content is evaluated as above by addition of a known amount of dimethylglyoxime using the addition method.

Nioxime

(Cyclohexane-1,2-dione dioxime)

Molecular weight: 142.156
White crystalline material. Melting point 184–189°C.
Solubility: 0.8% in water, readily soluble in alcohol, diethyl ether and acetone.

Nioxime is an analogue of dimethylglyoxime, with which, in general, the same determinations can be accomplished as with dimethylglyoxime. It is, however, more soluble in water than dimethylglyoxime so that it can be applied as an aqueous solution for the precipitation of metal complexes. In this way the potential error caused by the increased solubility of the complex in the presence of alcohol is eliminated. In gravimetric determinations the higher molecular weight of nioxime presents a further advantage (the molecular weight of the nioxime complex of nickel is 340.99, while that of the nickel complex of dimethylglyoxime is 288.91).

Recently nioxime has been successfully applied for the gravimetric determination of bismuth.[34, 477] The bismuth nioxime complex was precipitated selectively from a solution containing about 30 cations (among them nickel and palladium) using the masking action of EDTA at about pH 12.

α-Furyldioxime

Molecular weight: 220.18
White needles. Melting point 166–168°C.
Solubility: about 2% in water, readily soluble in ethanol and diethyl ether.

The chemical behaviour of furyldioxime is analogous to that of dimethylglyoxime. This reagent, too, possesses the advantageous feature of satisfactory solubility in water; this permits its application in aqueous solution and thus the elimination of the undesired effects of the presence of ethanol, discussed in connection with dimethylglyoxime.

A further advantage of the reagent appears in the combined solvent extraction–spectrophotometric determinations: the absorption maxima of both the nickel and the palladium complexes of this ligand fall (in chloroform solution) in the visible region. The absorption maximum of the nickel complex is at 435 nm with molar extinction coefficient 1.9×10^4,[340] while that of the palladium complex is at 436 nm with molar extinction coefficient 2.0×10^3.[342]

Salicylaldoxime

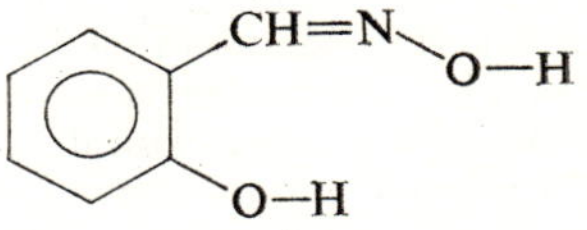

Molecular weight: 137.14
White crystalline powder. Melting point 57°C.
Solubility: sparingly soluble in water, soluble in ethanol, diethyl ether, acetone, benzene.

Salicylaldoxime reacts with several metal ions to give intensively coloured complexes which are insoluble in water. The structure of the copper(II), nickel(II) and palladium(II) complexes is shown in Fig. 14 (structure I), and that of the manganese(II), iron(II), cobalt(II) and zinc(II) complexes in Fig. 14 (structure II).[99] The copper(II) and palladium(II) salicylaldoxime complexes are of importance from an analytical point of view. The complexes precipitated from aqueous media are easily filterable and washable and are weighed after drying at 100°C. The precipitates are of stoichiometric composition.[167]

The outstanding stabilities of the copper(II) and palladium(II) complexes make the reagent suitable for the selective determination of these metal ions.[210] These two complexes form quantitatively at low pH values where the complexes of other metal ions do not exist.

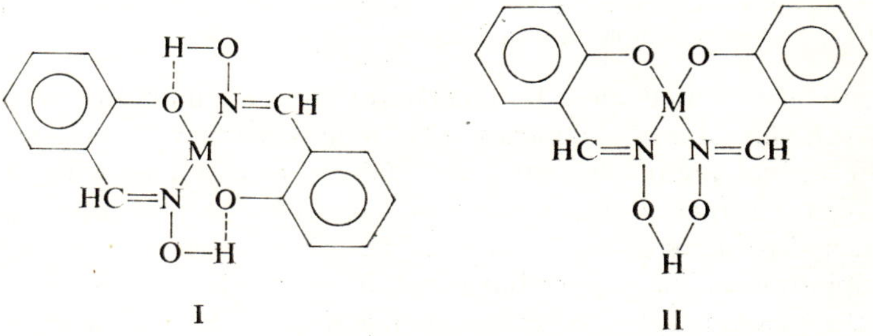

Fig. 14. Structural formulae of salicylaldoxime chelates: (I) complexes with copper(II), nickel(II) and palladium(II); (II) complexes with manganese(II), iron(II), cobalt(II) and zinc(II)

Besides the gravimetric method, salicylaldoxime is also useful in the volumetric, amperometric and nephelometric determinations of copper and palladium.

Salicylaldoxime gives a black precipitate with VO_3^- in media containing sulfuric acid; this is soluble in chloroform to give an orange solution. The reaction is specific for vanadium which can thus be detected by this method in the presence of almost all other metals.

In neutral medium salicylaldoxime reacts with iron(III) to give a red complex which is soluble in water. The reaction is utilized for the detection of micro amounts of iron.

Preparation of the reagent solution. 1 g salicylaldoxime is dissolved in 5 ml ethanol and the solution is added dropwise with continuous stirring to 95 ml hot (70–80°C) water. After cooling the solution is filtered. Since salicylaldoxime hydrolyses on standing in aqueous solution, a freshly prepared solution is always applied for quantitative analytical purposes.

Determination of copper. The sample solution containing the copper to be determined is neutralized with 2 M sodium hydroxide solution until precipitation commences and the precipitate is dissolved by addition of a few milliliters of 2 M acetic acid solution. The salicylaldoxime complex is precipitated from this solution (pH 3–5) at room temperature. After 30 minutes the yellow-green precipitate is collected on a sintered glass filter and washed with cold water until the washings do not give a colour with iron(III). The product is dried at not higher than 105°C and weighed. The gravimetric multiplication factor is 0.1895 for copper.

In the presence of iron tartaric acid is added to the solution as masking agent.

Determination of palladium. The palladium(II) salicylaldoxime complex precipitates quantitatively even in the presence of strong mineral acids (at about pH 2). In the case of palladium(IV) salts the precipitation is carried out in hot solution when the salicylaldoxime reduces the palladium(IV) and the precipitate contains only palladium(II). The precipitate is washed with cold water in both cases, dried at 110°C and weighed. (In the determination of palladium(IV) it is advisable to wash the precipitate with a few ml of 30% ethanol too, in order to remove the oxidation products formed in the reduction of palladium(IV).) The gravimetric multiplication factor is 0.2817 for palladium.

α-Benzoinoxime

Molecular weight: 227.25
White crystalline material which turns grey on exposure to light. Melting point 149–151°C.
Solubility: sparingly soluble in water, readily soluble in ethanol.

α-Benzoinoxime is used most frequently for the determination of copper(II), but it gives highly stable complexes with molybdenum(VI), tungsten(VI) and vanadium(V), which are insoluble in water and can be extracted with chloroform.

Copper can be quantitatively precipitated from hot solutions containing ammonium hydroxide or sodium tartrate. The excess reagent ensures the quantitative precipitation of the copper but it requires careful work to avoid contamination of the precipitate by the excess of the reagent. Hence combined solvent extraction–spectrophotometric measurements have proved more suitable than the gravimetric method. The absorption maximum of the copper chelate in chloroform solution is at 440 nm.[321]

Daxime

(1,3-Dimethylalloxane-imide(4)-oxime(5))

Molecular weight: 184.16
Violet crystalline material.
Solubility: sparingly soluble in water and acetone, practically insoluble in chloroform, alcohols, diethyl ether and dioxan, readily soluble in formamide. It dissolves in alkaline aqueous media with salt formation; the solution is reddish-brown.

The molecule contains two characteristic functional groupings:

1. the iso-nitroso-imino grouping specific for nickel and palladium(II);

2. the iso-nitroso-keto and nitroso-enol groupings which are in tautomeric

equilibrium and which are characteristic for iron(II), copper(II) and cobalt(II).

As one might expect from its structure, the reagent gives coloured complexes with copper(II), iron(II) and cobalt(II) which are soluble in water; with palladium and nickel ions, in slightly acidic and ammoniacal media, respectively, complexes which are practically insoluble in water are formed.

Daxime has proved suitable for the determination of micro amounts of copper(II) in the presence of large amounts of mercury(II), manganese(II), lead(II), zinc(II), cadmium(II), iron(III) and bismuth(III). The reagent gives a grass-green mixed complex (absorption maximum at 382 nm) with copper(II) in ammoniacal medium (Fig. 15). The molar extinction coefficient of the complex is 5050 at the absorption maximum.

Preparation of the reagent solution. A 0.05 M solution of Daxime (about 1%) is used.
Determination of copper. The sample solution containing 10–130 γ copper is concentrated to about 1–2 ml and approximately neutralized with 2 M base or acid using universal indicator paper to check the pH of the solution. If necessary, the solution is again concentrated to 1–2 ml. Subsequently 0.4 ml 1% Daxime solution and then 5–6 ml buffer solution are added. Whenever the solution contains ions which form precipitates with the buffer, the procedure is

carried out in a centrifuge tube and the precipitate is separated by centrifuging. After removal of the supernatant solution the precipitate is mixed with 1–2 ml water and the mixture again centrifuged. The two supernatant solutions are combined, the pH is adjusted to 7–9 if necessary, and the solution diluted to 10.0 ml. The extinction of the solution is measured at 382 nm in a 1 cm cell.

Multiplication of the measured extinction value by 12.6 gives the concentration of copper in γ/ml units.

In the case of greater amounts of copper the procedure is carried out in greater volume (25, 50 or 100 ml instead of 10 ml). Of course, the amounts of reagent and buffer are also increased proportionally.[83]

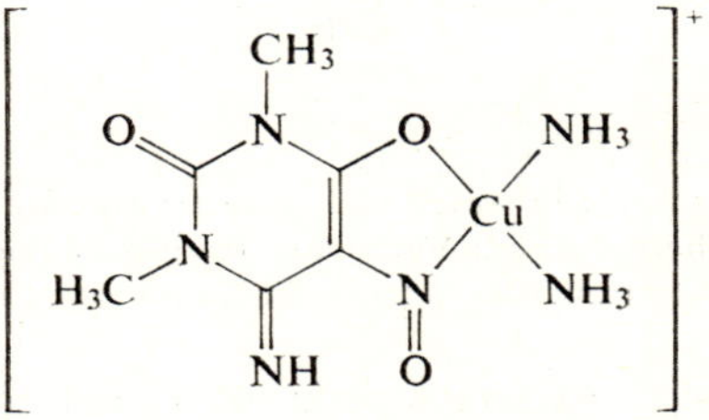

FIG. 15. The mixed complex of copper(II) formed with Daxime in ammoniacal medium

The analytical applicability of other ligands containing, as does Daxime, more than one type of functional grouping has also been investigated. Belcher *et al.*[41] have prepared several pyridyloxime ligands. They found that these are formed as mixtures of isomers which can be separated by thin-layer chromatography. These researches are of primary importance, particularly because they reveal that different isomeric modifications of organic molecules react with metal ions in different manners.

3. NITROSO COMPOUNDS

α-Nitroso-β-naphthol

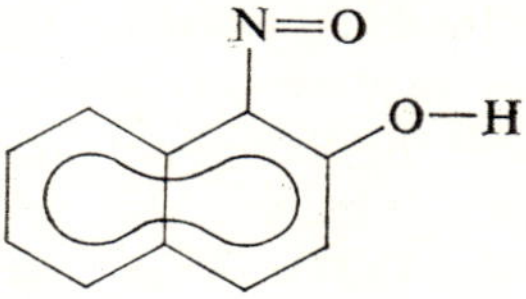

Molecular weight: 173.17
Orange crystals or brown powder. Melting point 108–110°C.
Solubility: it dissolves in 1000 parts of water or 35 parts of ethanol; readily soluble in benzene, diethyl ether, carbon disulfide and acetic acid. It is also soluble in alkaline aqueous media.

The reagent gives a chelate complex of composition MA_3 with cobalt(III). It also reacts with cobalt(II), but in this case the reagent oxidizes the metal ion in the first step and therefore the resulting cobalt(III) complex is contaminated by the reduction products of the reagent. Consequently, when cobalt(II) is to be determined it is transformed first to cobalt(III) hydroxide by treatment with hydrogen peroxide in alkaline solution. The product is dissolved in acetic acid to give cobalt(III) acetate which reacts cleanly with α-nitroso-β-naphthol.

The cobalt(III) complex is formed in slightly acidic, neutral or ammoniacal media, but due to its inert character it does not dissolve in mineral acids. With this reagent cobalt can be determined in the presence of nickel.[291] The analytical selectivity of the ligand was dealt with in Chapter 1, p. 57.

Preparation of the reagent solution. 4 g α-nitroso-β-naphthol is dissolved in 100 ml glacial acetic acid (in order to facilitate dissolution, the reagent is first triturated with a little glacial acetic acid). 100 ml hot water is added to the glacial acetic acid solution and the whole is filtered.

Gravimetric determination of cobalt. 100 ml 6 N hydrochloric acid is added to the approximately neutral sample solution containing about 0.1, but a maximum of 0.2 g cobalt and the solution is diluted to 200 ml. The reagent solution is added in excess to the hot (80°C) sample solution. The mixture is allowed to stand 2–3 hours and filtered. The precipitate is washed with a few ml 12% hydrochloric acid and then with hot water, followed by ignition in a stream of hydrogen. The resulting metallic cobalt is weighed.

Solvent extraction determination of cobalt. Micro amounts of cobalt can be determined selectively in samples of biological origin (e.g. in bones) by means of the following solvent extraction method.

The sample is moistened with concentrated nitric acid and carefully heated at about 400° until completely decomposed. The residue is dissolved in 25 ml M phosphoric acid. The solution is neutralized with 10% sodium citrate solution in the presence of methyl orange indicator. 5 ml 1% 1-nitroso-2-naphthol solution (in 50% acetic acid) is added. The mixture is kept for 1 hour with occasional shaking and extracted three times with 10 ml chloroform.

β-Nitroso-α-naphthol

Molecular weight: 173.17
Yellow crystalline material. Melting point 147–148°C.
Solubility: sparingly soluble in cold water, readily soluble in ethanol.

β-Nitroso-α-naphthol behaves similarly to α-nitroso-β-naphthol. In practice it has no advantage over α-nitroso-β-naphthol. Hence a detailed discussion of the reagent is unnecessary.

Cupferron

(Ammonium salt of N-nitroso-phenylhydroxylamine)

Molecular weight: 169.16
White crystalline material. Melting point 163–164°C.
Solubility: well soluble in water and ethanol.

The reagent used to be considered as specific for iron(II) and copper(II). However, in the course of later analytical studies it turned out that cupferron forms complexes with several other metals too, even in acidic medium. It is used today primarily for the determination of copper(II), but it has proved suitable for the determination of aluminium(III), bismuth(III), iron(III), mercury(II), thorium(IV), tin(IV), titanium(III), vanadium(IV) and zirconium(IV) as well. Of these, the zirconium(IV) complex is the most stable. Cupferron precipitates zirconium(IV) quantitatively from aqueous media containing sulfuric acid.

The reagent gives a precipitate with most of the above metal ions even in strong mineral acidic medium. Only the aluminium(III) complex does not precipitate in the presence of mineral acids.

By means of cupferron many metal ions can be separated: thus, for instance, iron(III) can be separated from aluminium(III) and manganese(II) and titanium(IV), zirconium(IV) and hafnium(IV) from many other metal ions, etc. Though the reagent makes possible several separations, the precipitates are not exactly of stoichiometric composition (they almost always contain some excess of the ligand). Hence instead of being directly weighed, the precipitates are usually ignited and the metal oxide residues weighed.[435]

4. NITRO COMPOUNDS

α-Nitro-β-naphthol

Molecular weight: 189.17

The reagent is useful for the detection and determination of cobalt under conditions similar to those used with α-nitroso-β-naphthol. By means of α-nitro-β-naphthol cobalt can be determined in the presence of nickel, zinc, manganese, aluminium, chromium and iron, while palladium and platinum can be separated.[290]

An important advantage of the reagent is its enhanced resistance towards oxidizing agents as compared to the corresponding nitroso compound. Hence its solutions are much more stable and can be stored for longer periods.

The reagent is easily prepared by treatment of α-nitroso-β-naphthol with nitric acid.

Picrolonic acid

(1-(p-nitrophenyl)-3-methyl-4-nitro-5-pyrazolone)

Molecular weight: 264.19
Yellow crystalline material. Melting point 116–120°C (with decomposition).
Solubility: 0.27% in water at room temperature, readily soluble in alcohol.

With calcium(II), lead(II) and thorium(IV) the reagent gives complexes which are practically insoluble in water. The high molecular weight crystalline complexes can be used to advantage for the accurate determination of minute amounts of various metals.

Preparation of the reagent solution. 2.64 g finely powdered picrolonic acid is dissolved in 100 ml distilled water with shaking. The reagent solution can be stored for long periods without decomposition.

Determination of calcium. The sample solution containing 0.1 g calcium is heated to 40–50°C and the reagent solution is added dropwise with continuous stirring until a drop causes no further precipitation. The reaction mixture is heated to boiling in order to recrystallize the precipitate. After cooling, a further volume of reagent solution is added (half the amount initially added). The mixture is kept for several hours and the precipitate is collected on a sintered glass filter, washed with 4×2 ml cold water and dried either by suction of air or in vacuum desiccator. The gravimetric multiplication factor for calcium is 0.05641.[324]

The reagent has also proved suitable for the spectrophotometric determination of calcium. Calcium picrolonate is transformed by bromine water in alkaline medium to a red, water-soluble complex. The absorption of this compound is constant for about two days and it also follows the Beer–Lambert Law.[250]

5. OXINE AND ITS DERIVATIVES

8-Hydroxyquinoline (Oxine)

Molecular weight: 145.15
White crystalline compound. Melting point 74–76°C, boiling point 267°C.
Solubility: practically insoluble in water and diethyl ether, easily soluble in ethanol, acetone, chloroform, benzene and mineral acids.

Oxine is an almost universal complexing agent; it reacts with very many metal ions to give water-insoluble precipitates. It has been used so far for the determination of aluminium, antimony, beryllium, bismuth, cadmium, cerium, cobalt, chromium, copper, gallium, germanium, lanthanum, lithium, magnesium, manganese, molybdenum, nickel, ruthenium, thorium, titanium, uranium, vanadium, tungsten, zinc and zirconium.[49]

With the exception of the aluminium, bismuth, gallium, indium, lead and thallium complexes, the oxinate precipitates contain water of crystallization. The majority of the complexes are of stoichiometric composition (either with or without water of crystallization), and therefore subsequent to drying at suitable temperature the precipitates are ready for direct weighing. The precipitates containing water of crystallization are usually dried at 100–105°C and weighed as such, or dried at 130–140°C when the water of crystallization is eliminated.

The complexes can be decomposed with strong acids, and the oxine liberated in an amount equivalent to the metal may be determined by bromatometric titration.

The electrically neutral oxinate complexes are sparingly soluble in water whereas they readily dissolve in apolar solvents immiscible with water, for instance chloroform. Hence this ligand is useful for the solvent extraction enrichment and the subsequent spectrophotometric determination of metal ions (cf. Table 17).

The optimum pH range for complex formation is dependent upon the stabilities of the oxinate complexes (Table 18). Accordingly, divalent metals are usually precipitated quantitatively in neutral or slightly alkaline media. Adjustment of the suitable pH may be made with various tartrate buffers, for example with ammonia–ammonium tartrate in the pH range 6–10, and with sodium hydroxide–sodium tartrate buffers for the pH range 10–13. Complexes of tri- and tetravalent metals are more stable, and thus can be precipitated quantitatively even at about pH 4. In these cases the pH of the solution is adjusted to the desired value with acetic acid–sodium acetate buffers. The copper(II) and iron(III) complexes precipitate quantitatively even at pH 2.7, as does vanadium(V) as an oxinate complex of uncertain composition.

Preparation of the reagent solution. A 2 or 4% solution of oxine in ethanol or acetone is usually used. The solution must be prepared freshly before use; it can be stored without decomposition for a maximum of 8–10 days only. In cases when ethanol or acetone would considerably increase the solubility of the oxinate complexes to be formed, oxine acetate solution is used. 3–4 g oxine is dissolved in a few millilitres of glacial acetic acid and the solution diluted with distilled water to 100 ml. The solution is then neutralized with concentrated ammonium hydroxide to the first turbidity, and subsequently treated dropwise with 2 M acetic acid solution until a clear solution results. The solution can be stored for long periods without decomposition.

Analytical procedures. Various gravimetric, titrimetric and spectrophotometric analytical procedures based on oxine have been elaborated for the determination of different metal ions. The analytical procedure varies slightly in the measurements of the individual metal ions, depending upon the physicochemical properties of the complex concerned. The following general guiding principles are useful in devising the individual procedures:

1. Quantitative formation of the complex is ensured by excess of the reagent. In acidic media containing acetic acid the presence of excess reagent is indicated by the yellow colour of the solution, and in alkaline media by its orange colour. However, because of the slight solubility of the reagent in water, it should not be applied in great excess, and therefore the change in colour of the solution indicating the appearance of excess reagent should be carefully watched during precipitation.

2. Recrystallization of a complex is facilitated by heating the mixture at 60–70°C for a few minutes after precipitation.

3. In general it is advisable to filter the precipitate at high temperature. The filtrate may deposit reagent on cooling, but this redissolves on heating. Whenever the precipitate does not dissolve on heating, it contains metal ions, and in this case precipitation must be repeated with a greater excess of the reagent.

4. The precipitate collected on the filter is washed with hot water until the colour of the reagent is not detectable in the washings.

5. In gravimetric methods the precipitate is usually dried at 130–140°C. After drying at such temperatures, the complex is anhydrous even if it originally precipitated in a form containing water of crystallization.

6. Oxinate complexes are quantitatively soluble in 10% hydrochloric acid. The oxine solute, which is equivalent to the metal to be determined, can be titrated bromatometrically in the hydrochloric acid solution. Two bromine atoms are substituted into each oxine molecule and, accordingly, in the case of MA_2 complexes the equivalent weight of the central atom is one-eighth of the atomic weight, while with MA_3 complexes it is one-twelfth.

7. Oxine complexes are usually intensely coloured in chloroform. This renders possible the spectrophotometric determination of many metal ions. Thus, for instance, copper can be extracted with oxine into chloroform at pH 4 (even in the presence of fortyfold amounts of cadmium or zinc), and its concentration determined by measuring the absorption of the chloroform solution at 410 nm. Table 17 summarizes several selective solvent-extraction analytical procedures. The data of this Table illustrate well the analytical importance of oxine.

The analytical selectivity of the formation of oxine complexes may be further increased by combination of the solvent-extraction technique with chromatographic separation. Krausz,[254] for instance, has elaborated such a combined method for the selective determination of vanadium(V).

TABLE 17. APPLICATION OF 8-HYDROXYQUINOLINE (OXINE) FOR DETERMINATION OF METAL IONS BY SPECTROPHOTOMETRY COMBINED WITH SOLVENT EXTRACTION

Metal ion	pH of the aqueous phase	Extracting solvent	Absorption maximum, nm	Extraction constant,† $\log K$	Interfering ions	Successful application	Note, masking agent	Refs.
Al(III)	4.5–11	$CHCl_3$	390	−5.22	high concn. of metal ions	analysis of metals, plastics, etc.	CN^- masks Co, Ni, Zn, Cd, and Fe(II) ions	15, 19, 119, 188, 189, 382
Ba(II)	>10	$CHCl_3$	380–400	−20.9	high concn. of metal ions		repeated extraction is required. $BaA_2(HA)_2$ is found in the organic phase	442, 462
Be(II)	6–10	$CHCl_3$	380	−9.62	high concn. of metal ions	analysis of alloys		442
Bi(III)	2.5–11	$CHCl_3$	390–395	−1.2				442
Ca(II)	>10.7	$CHCl_3$	380–400				CaA_2HA gets into the organic phase	442
Cd(II)	5.5–9.5	$CHCl_3$	380–390	−5.29			$CdA_2(HA)_2$ in the organic phase	442, 461, 470
Ce(III)	9.9–10.6	$CHCl_3$	495–500		Fe(III)		the presence of citrate or tartrate is necessary	8

Ion	pH	Solvent	λ		Masking agents	Application	Remarks	References
Co(II)	4.5–10.5	$CHCl_3$	420	−2.16				442
Cr(III)	6–8	$CHCl_3$	425			the extraction may be specific for chromium due to the inert nature of the complex	formed at boiling temperature only	302, 442
Cu(II)	2–12	$CHCl_3$	410–420	1.4	CN^-, EDTA	determination of Cu in uranium ores		189, 302, 316, 442
Fe(III)	2–10	$CHCl_3$	470, 580	4.11	V(V), Ru(III), Ce(IV)	determination of the iron content of alloys		121, 206, 442
Ga(III)	2.2–12	$CHCl_3$	400 ($\varepsilon = 6470$)	3.72		analysis of ores and alloys		442
Hg(II)	>3	$CHCl_3$	390 ($\varepsilon = 5400$)					461
In(III)	3–11.5	$CHCl_3$	395–400 ($\varepsilon = 6670$)	0.89		separation of In and Be; analysis of ores and minerals	strong fluorescence	301, 442
La(III)	7–10	$CHCl_3$	380	−15.66				442
Mg(II)	~9	$CHCl_3$	385 ($\varepsilon = 5300$)	−15.13	high concn. of metal ions	analysis of minerals and alloys after separation	shaking longer than one minute causes decomposition	272–3, 442, 492

Table 17 (continued)

Metal ion	pH of the aqueous phase	Extracting solvent	Absorption maximum, nm	Extraction constant,† $\log K$	Interfering ions	Successful application	Note, masking agent	Refs.
Mn(II)	6.5–10	$CHCl_3$	395	−9.32				189, 442
Ni(II)	4.0–10.0	$CHCl_3$	390	−2.18				442
Pb(II)	6–10	$CHCl_3$	400	−8.04				442
Pd(II)	0–10	$CHCl_3$	425–430	15				231, 442
Sr(II)	>11.5	$CHCl_3$	380–400	−19.71			$SrA_2(HA)_2$ in the organic phase	442
Th(IV)	4–10	$CHCl_3$	380	−7.18			the Beer–Lambert Law is not followed	139, 442
Ti(IV)	2.5–9.0	$CHCl_3$	380	−0.90			$TiOA_2$ in the organic phase	121, 442
Ti(IV)	3–5.1% H_2O_2	$CHCl_3$	425					187
Tl(III)	3.5–11.5	$CHCl_3$	400	5				8, 121, 139, 187, 206, 231, 272–73, 299, 301–03, 315, 442, 462

U(VI)	5–9	CHCl$_3$	425–430	−1.60	interfering ions can be masked with EDTA	UO$_2$A$_2$HA in the organic phase	7, 8, 121, 124, 139, 187, 206, 231, 272, 273, 299, 301–03, 315, 442, 462
V(V)	2–6	CHCl$_3$	380, 580	1.67		analysis of rock	VO$_2$A in the organic phase	395, 442
Zn(II)	4–5	CHCl$_3$	380–400				ZnA$_2$(HA)$_2$ in the organic phase	442
Zn(II)	10–11	CHCl$_3$ + n-butyl-amine	370 (ε = 5210)				ZnA$_2$H$_2$O and butylamine in the organic phase	461
Zr(IV)	1.5–4.0	CHCl$_3$	385 (ε = 14,000)	2.71		analysis of alloys after separation	ZrOA$_2$ in the organic phase	316

†$K = \dfrac{[MA_N]_{org}[H]^N}{[M][HA]^N_{org}}$, where [MA$_N$]$_{org}$ = concentration of the oxinate complex in the organic phase; [HA]$_{org}$ = concentration of oxine in the organic phase.

TABLE 18. pH-RANGES OF THE QUANTITATIVE DEPOSITION OF STOICHIOMETRIC METAL OXINATES

Central atom	pH range	Central atom	pH range
Al(III)	4.2– 9.8	Mn(II)	5.9– 9.5
Cd(II)	5.7–14.6	Mo(VI)	3.6– 7.3
Cu(II)	2.7–14.6	Pd(II)	3.5– 8.5
Fe(III)	2.8–11.2	Sc(III)	6.5– 8.5
Ga(III)	3.6–11	U(VI)	5.7– 9.8
In(III)	2.5– 3.0	W(VI)	5.0– 5.7
Mg(II)	9.6–12.7	Zn(II)	4.7–13.5

Some metal oxinates, e.g. those of aluminium, gallium, indium and zinc, exhibit strong fluorescence in chloroform solution. This feature may serve as the basis of the fluorometric determination of these metals.

In addition to the above-detailed direct analytical procedures, some oxinate complexes can also be determined indirectly.[13, 47, 190] Thus, for instance, decomposition of the oxinate complex of a divalent metal ion, subsequent addition of iron(III), and the spectrophotometric measurement of the absorption of the iron(III) oxinate complex produced even in quite acidic media, renders possible the determination of the divalent metal ion. Treatment with nitrous acid results in diazotization of the hydroxyquinoline ligand, and coupling with some suitable reagent gives rise to an azo compound, whose colour also may be utilized for colorimetric determination.

Detailed descriptions of some of the more important analytical procedures elaborated from the above general principles are given in the following:

Gravimetric determination of aluminium. The aluminium oxinate complex can be precipitated from both dilute acetic acid medium and ammoniacal ammonium tartrate solution. In the presence of trivalent metal ions the analytical selectivity of the precipitation is higher in the latter medium. Thus, for instance, it permits the separation of aluminium(III) from iron(III) and beryllium(III). On the other hand, in the presence of divalent metal ions, precipitation from acetic acid solution possesses the higher selectivity. A procedure checked by the author is given for the latter type of precipitation:

Preparation of the reagent solution. 5 g oxine is dissolved in 12 g glacial acetic acid and the solution is diluted with distilled water to 100 ml.

Procedure. 100 ml sample solution containing about 0.05 g aluminium(III) is acidified with 10 ml reagent acetic acid, heated to 70°C and treated with 30 ml acetic acid oxine reagent solution. 15% ammonium acetate solution is then added with continuous stirring until the precipitate is deposited, followed by a further 25 ml of the same solution. The mixture is kept for an hour, filtered, and the precipitate washed with cold water and dried at 120°C. Gravimetric factor: 0.05872.

Solvent-extraction determination of aluminium. Aluminium can be determined selectively in the presence of copper, nickel, cobalt, zinc, cadmium, tungsten and molybdenum by extraction with a solution of oxine in chloroform.

Analytical procedure. About 50 ml sample solution containing 10–50 µg/ml aluminium is treated with 2 M ammonia solution to about pH 9, and 1 g potassium cyanide is dissolved in the alkaline solution. The aqueous phase is then extracted with 10 ml of a 1% solution of oxine in chloroform. The absorption maximum of the aluminium complex appears at 395 nm in chloroform solution.

Gravimetric determination of cobalt. About 100 ml neutral sample solution containing 10–15 mg cobalt is treated with 3 g crystalline sodium acetate, and then with 10 ml 10% acetic acid solution. The complex is precipitated by addition of 2% oxine solution in ethanol at about 70°C. The reagent is added until the supernatant solution is pale yellow, indicating the presence of a slight excess unreacted oxine. The precipitate is light-brown, but the colour turns to flesh-pink on heating the mixture for 15–20 minutes on a water bath. The precipitate is collected on a filter, washed with warm water and dried at 180°C. Gravimetric factor: 0.1697.

Solvent-extraction determination of cobalt. The following solvent-extraction method has proved suitable for the selective determination of cobalt in the presence of iron, copper and bismuth.

These accompanying ions can be removed quantitatively from aqueous solutions of cobalt at pH 3.5 by extraction with a chloroform solution of oxine. Under such acidic conditions the cobalt complex is not formed; its quantitative formation requires raising the pH, after removal of the accompanying ions, to above 7. Subsequently, the cobalt can be extracted with a 1% oxine solution in chloroform. Aluminium(III) and nickel(II) interfere with the determination.

Gravimetric determination of copper. The copper oxinate complex precipitates from both dilute acetic acid and alkaline sodium tartrate solutions. The dilute acetic acid medium ensures the selective determination of copper in the presence of beryllium, magnesium, calcium, cadmium, lead, arsenic and manganese. On the other hand, precipitation of the copper complex occurs selectively in the presence of aluminium, lead, tin(IV), arsenic(V), antimony(V), bismuth, chromium and iron(III) in alkaline tartrate medium.

Precipitation from acetic acid solution. 3 g crystalline sodium acetate and 10–12 ml glacial acetic acid are dissolved in about 80 ml neutral sample solution containing 10–80 mg copper. The solution is heated to about 60–70°C and 3% oxine solution in ethanol is added dropwise with stirring until the supernatant solution is pale yellow. The mixture is heated a few minutes on a boiling water bath, filtered through a G4 sintered glass filter, and the precipitate washed with hot water and dried at 105°C. Gravimetric factor: 0.1806.

Precipitation from alkaline sodium tartrate solution. About 80 ml neutral or slightly acidic sample solution containing 10–80 mg copper is treated with 5 g tartaric acid and neutralized with 2 M sodium hydroxide solution in the presence of phenolphthalein indicator. After addition of a further 20 ml 2 M sodium hydroxide solution the copper oxinate complex is precipitated with a 3% solution of oxine in acetone. The appearance of excess reagent is indicated by its orange colour. (In the presence of iron(III) the appearance of excess reagent is indicated by a dark brown colour.) The mixture is heated to about 70°C and allowed to cool, the precipitate collected on a G4 sintered glass filter, washed with 1% sodium tartrate solution and a few ml of cold water, and dried at 105°C.

Solvent-extraction determination of copper. Up to 300 µg copper can be selectively determined in the presence of iron, aluminium, cobalt, nickel and manganese by the following solvent extraction procedure.

Analytical procedure. About 50 ml sample solution is treated with 5 ml 0.02 M EDTA solution and the pH is adjusted to 6.5. The solution is then extracted with 2 × 10 ml 1% oxine solution in chloroform. The absorption of the complex is measured at 400 nm in chloroform solution. Uranium, vanadium, molybdenum and titanium interfere with the determination

Gravimetric determination of magnesium. The magnesium oxinate complex forms only in solutions of pH > 7. Thus, all those metals which give oxinate complexes in dilute acetic acid can be selectively precipitated in the presence of magnesium(II). Subsequent treatment of the filtrate with ammonium hydroxide gives the magnesium oxinate complex. Only calcium, gallium and beryllium interfere with this determination of magnesium, for the oxinates of these metals also form only in alkaline medium.

Procedure. After removal of the accompanying ions in acetic acid solution, the sample solution containing 10–50 mg magnesium is treated with concentrated ammonia solution to about pH 10. In order to keep the magnesium in solution, 2 g ammonium chloride is added to the solution before the treatment with ammonia. The mixture is then heated to about 70°C and 2% oxine solution in ethanol is added dropwise with stirring until the supernatant solution is yellow. The mixture is heated for 20–30 minutes on a boiling water bath, and the precipitate is collected on a G3 sintered glass filter and washed with hot 1% aqueous ammonia solution. The optimum drying temperature is 160°C. Gravimetric factor: 0.0778.

Gravimetric determination of molybdenum. At pH 3.5–7.5 hexavalent molybdenum gives an orange oxinate complex of composition $MoO_2(C_9H_6NO)_2$ which is practically insoluble in the usual solvents and dilute mineral acids.

Analytical procedure. About 50 ml sample solution containing 5–50 mg molybdenum is neutralized in the presence of methyl red indicator and acidified with 10 ml 2 M acetic acid. After addition of 10 ml M ammonium acetate solution, the mixture is heated to boiling, the gas burner is removed and 3% oxine solution in acetic acid is added.
 The precipitating agent should be applied in small excess. The appearance of excess reagent is indicated by the yellow colour of the supernatant solution. After precipitation the mixture is heated again to boiling and filtered through a G4 sintered glass filter while hot. The precipitate is washed with hot water and dried at 130–150°C. Gravimetric factor: 0.2305.

Solvent extraction determination of molybdenum. The oxinate complex of molybdenum can be extracted with chloroform at pH $\geq$ 1.4. Metal ions which interfere even under such acidic conditions can be masked with EDTA.

Procedure. The sample solution of not more than 100 ml containing up to 500 μg molybdenum is treated with 5–10 ml 0.02 M EDTA solution, the pH of the solution is adjusted to 1.5–1.6, and the aqueous phase is extracted twice with 10 ml 1% oxine solution in chloroform. The absorption of the complex is measured at 370 nm in chloroform solution. Tungsten interferes with the determination.

Bromatometric determination of metal oxinates. A metal precipitated in the form of its water-insoluble oxinate complex can also be determined volumetrically by dissolution of the complex precipitate in hydrochloric acid and bromatometric measurement of the oxine content, which is equivalent to the metal.

Procedure. The metal oxinate precipitate, obtained according to the gravimetric analytical prescription and washed thoroughly with water, is dissolved in 20% hydrochloric acid. The solution is then diluted with water so that 50 ml diluted solution contains 0.03–0.06 g hydroxyquinoline and the acid concentration is 2 M. After dissolution of 1 g potassium bromide in the solution, the oxine content is titrated with 0.1 N potassium bromate standard solution in the presence of 2 drops of *p*-ethoxychrysoidine indicator. After decoloration of the indicator a further 0.1–0.2 ml standard solution is added. The excess potassium bromate is titrated iodometrically after 2–3 minutes. Prior to titration with thiosulfate of the iodine produced, the solution is diluted to 150 ml in order to eliminate the interfering effect of the formation of a "periodide" precipitate. 1 ml 0.1 N potassium bromate standard solution is equivalent to 3.629 mg oxine.

8-Hydroxyquinaldine

(2-methyl-8-hydroxyquinoline)

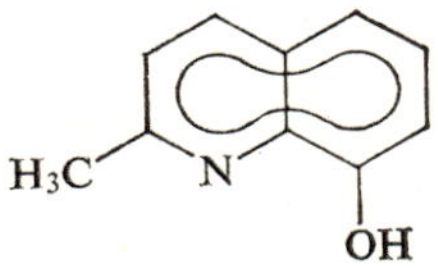

Molecular weight: 159.17
White or pale yellow compound, melting point: 74°C.
Solubility: sparingly soluble in water, freely soluble in diethyl ether, chloroform, benzene. In aqueous solution it acts as a weak acid, with acidic dissociation exponent 10.04.

This compound possesses analytical features analogous to those of oxine. However, due to the steric hindrance arising from the methyl group in the ortho position to the donor nitrogen atom, its analytical selectivity is higher than that of oxine. For instance, it does not form a precipitate or a complex extractable by apolar solvents with aluminium(III).[295] Many metal ions can therefore be separated from aluminium by means of this reagent.[214]

6. PAN

(1-(2-pyridylazo)-2-naphthol)

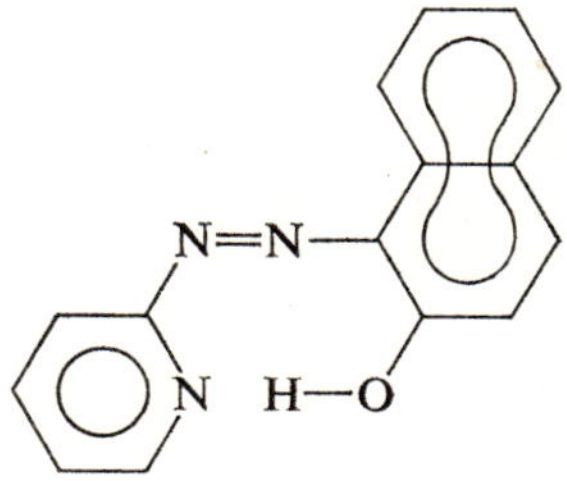

Molecular weight: 249.28
Orange amorphous substance.
Solubility: practically insoluble in water, dissolves in alkali with salt formation. Soluble in organic solvents to give yellow solutions, with absorption maximum at 470 nm. The reagent does not absorb above 560 nm.

PAN reacts with many metal ions to give intensely coloured chelate complexes, which can be extracted with carbon tetrachloride, chloroform, benzene or diethyl ether.[117, 426] Its palladium(II) and cobalt(III) complexes are green, while the other metal complexes are various shades of red.

In addition to dithizone, oxine, diethyl dithiocarbamate and acetylacetone, PAN is one of the most important extraction reagents in analytical chemistry. Its analytical applications are summarized in Table 19.

Preparation of the reagent solution. For analytical purposes 0.1% solutions of PAN in ethanol or methanol are used.

TABLE 19. APPLICATION OF PAN FOR DETERMINATION OF METAL IONS BY SPECTROPHOTOMETRY COMBINED WITH SOLVENT EXTRACTION

Metal ion	pH of the aqueous phase	Solvent	Absorption maximum, nm	Molar extinction coefficient	Successful application	Note	Refs.
Cd(II)	7–10	CHCl$_3$	550–560	49,000–51,000	determination of Cd in Ni		52, 427
Co(II)	4–7	CHCl$_3$					52
Co(III)	3–6	CHCl$_3$	590 640	25,000 20,000	determination of Co in thorium oxide		195
Cu(II)	4–10	CHCl$_3$	550	45,000		extractable with amyl alcohols	52, 117
Fe(III)	4–7	CHCl$_3$ or benzene	775	16,000	mineral analysis		52, 426, 427
Ga(III)	6–7.5	CHCl$_3$	560				427
Hg(II)	6–7.5	CHCl$_3$	560				427
In(III)	5.3–6.7	CHCl$_3$	560	19,000			426, 427
Mn(II)	7–10	CHCl$_3$	550	40,000			52

Ni(II)	4–10	CHCl$_3$ or benzene	575	50,900		formed quantitatively at 80°C	424
Os(VIII)	7.8–9.5	CHCl$_3$	560–570	11,300			106
Pd(II)	3–7	CHCl$_3$	675–678	14,000–16,000	determination of Pd in alloys	Rh, Pt, Au, Ag, and Hg do not interfere in solution of pH 3–4	52, 107, 405
U(VI)	5–10	o-dichloro-benzene	570	23,000		EDTA and CN$^-$ may be used to mask interfering metal ions	114, 116
V(V)	3.5–4.5	CHCl$_3$	615	16,900	analysis of alloys and organic substances		445, 468
Y(III)	8.5–11	diethyl ether	560				427
Zn(II)	4.5–8	CHCl$_3$	550–560		determination of Zn in Ni		51, 52

7. ORGANIC ACIDS

Anthranilic acid

(*o*-Aminobenzoic acid)

Molecular weight: 137.13
White or pale yellow crystalline substance. Melting point 144–146°C.
Solubility: sparingly soluble in cold water, freely soluble in hot water, ethanol, diethy ether. Its solutions in organic solvents fluoresce.

The reagent is used for the detection and determination of cadmium(II), cobalt(II), copper(II), lead(II), manganese(II), mercury(II), nickel(II) and zinc(II).[183, 476] It usually gives metal complexes at pH 4.5–6.0. Only the copper complex forms at pH 2.8 (Table 20), and hence the reagent is useful for the

TABLE 20. pH-Dependence of Metal Anthranilate Precipitation

Central atom	Lower limit of precipitation	Lower limit of quantitative precipitation
Cd(II)	4.25	5.23
Co(II)	3.36	4.41
Cu(II)	1.40	2.79
Mn(II)	2.40	5.15
Ni(II)	3.64	4.51
Zn(II)	3.76	4.72

selective determination of copper in the presence of the above metals. The zinc, cobalt, cadmium and nickel complexes are slightly soluble in sodium acetate solution.

After drying at 105–110°C, the anthranilate precipitates are ready for direct weighing, but they can also be used for the oxidimetric determination of the anthranilic acid content, which is equivalent to the metal to be determined.[152, 204]

Preparation of the reagent solution. 3.0 g anthranilic acid is dissolved in 22 ml M sodium hydroxide solution. The filtered solution must be slightly acidic or neutral; its colour may be pale yellow. It is stored in a well-closed brown glass container protected from light. Under such conditions it can be stored for long periods. Browning of the solution indicates decomposition of the reagent.

Gravimetric determination of zinc. 100–150 ml sample solution containing about 100 mg zinc is treated with acetic acid to about pH 5, heated to boiling and 4 ml reagent solution is added dropwise at this temperature. After standing 10–15 minutes, the precipitate is collected on a sintered glass filter, washed with 0.1% reagent solution, dried at 105–110°C and weighed. Gravimetric factor: 0.1937.

The other metals mentioned can be determined analogously, with the exception that cadmium(II) and manganese(II) require pH 5.5 for quantitative precipitation, while copper(II) precipitates quantitatively even at pH 3.

Volumetric determination of zinc. The complex precipitate obtained as above is collected quantitatively on a paper filter, washed thoroughly and dissolved in about 25–30 ml 10% hydrochloric acid. The solution is cooled with ice-water, treated with 1–2 g potassium bromide and titrated with 0.1 N potassium bromate standard solution in the presence of 2 drops of *p*-ethoxychrysoidine indicator. The standard solution is added in slight excess (max. 0.5 ml), and the excess is immediately titrated with 0.1 N sodium thiosulfate solution, after addition of 0.1–0.2 g potassium iodide, in the presence of starch indicator. The thiosulfate consumption is deduced from the potassium bromate consumption. 1 ml 0.1 N bromate standard solution is equivalent to 0.8173 mg zinc.

Quinaldic acid

(Quinoline-2-carboxylic acid)

Molecular weight: 173.17
White, odourless crystalline powder. Melting point 155–157°C.
Solubility: sparingly soluble in water, well soluble in ethanol and alkali.

The reagent gives MA_2 chelate complexes, which usually contain water of crystallization also, with divalent ions such as copper, lead, cadmium, zinc, manganese, nickel, cobalt and iron. On the other hand, it gives basic complexes with trivalent metal ions. In the absence of iron and zinc, the reagent is specific for copper in dilute sulfuric acid medium, for the copper(II) complex, due to its greater stability, precipitates at a lower pH than the other related metal complexes.[372, 373, 490, 491] It is noteworthy that the formation of the complex is influenced not only by the pH of the solution but also by the anion of the acid present. For instance, the solubility of the copper complex in acetic acid solution at pH 2.05 is ten times as great as that in sulfuric acid solution at the same pH value. The phenomenon is explained by the complexing ability of the acetate ion.

Preparation of the reagent solution. The sodium salt of quinaldic acid is used as reagent (5.636 g sodium quinaldate is dissolved in 150 ml water). The reagent can be recovered after consumption by treatment of the filtrate with copper(II), washing the precipitate and decomposition of a water suspension of the complex with hydrogen sulfide. After removal of the copper sulfide precipitate by filtration, the solution is evaporated to dryness and the residue purified by crystallization from glacial acetic acid.

Gravimetric determination of copper. The sample solution containing 10–50 mg copper in about 150 ml, and which is about 0.05 N in sulfuric acid, is heated to boiling. The precipitating agent is added dropwise at boiling temperature. The pale green precipitate is collected on a sintered glass filter after a few minutes, washed with hot water and dried at 125°C. The precipitate contains 1 mole water of crystallization. The gravimetric factor is 0.1496.

The procedure makes possible the accurate determination of the copper content of steels in the presence of tartrate as masking agent. Tartrate does not interfere with the precipitation of the copper quinaldate complex, while it does mask all the other constituents of the steel. In a steel analysis, about 1 g sample is dissolved in a mixture of 20 ml 20% hydrochloric acid and 2–3 ml concentrated nitric acid. The solution is filtered to remove undissolved fractions (silica, etc.) and evaporated to dryness. The dry residue is treated with a mixture of hydrogen fluoride and concentrated sulfuric acid and the mixture again evaporared to dryness (removal of the last traces of silicates). The dry residue is dissolved in water containing a little sulfuric acid, 6 g tartaric acid is added and the solution neutralized with ammonium hydroxide. After addition of 15 ml 4 N sulfuric acid the copper quinaldate complex is precipitated as above.

By means of quinaldic acid copper can be determined in the presence of cadmium, lead, manganese, nickel and cobalt. Separation is rendered possible by the high stability of the copper complex, as a consequence of which the complex precipitates quantitatively in dilute sulfuric acid solution even at pH 2, while formation of the corresponding complexes of the above metals requires neutral conditions.

Mandelic acid

(α-Hydroxyphenylacetic acid)

Molecular weight: 152.14
White crystalline compound. Melting point 117–118°C.
Solubility: soluble in 6.3 parts water and in 1 part ethanol. Readily soluble in diethyl ether and isopropanol.

Mandelic acid is an approved reagent for the selective determination of zirconium. By means of this reagent zirconium can be determined accurately in the presence of several-fold amounts of titanium, iron, vanadium, aluminium, chromium, thorium, cerium, tin, barium, calcium, copper, bismuth, antimony and cadmium.[200, 201, 260]

The complex precipitate of stoichiometric composition MA_4 may serve as the basis for the gravimetric determination of zirconium. Another possibility is the dissolution of the complex precipitate in concentrated sulfuric acid and titrimetric measurement of the mandelic acid ligand by total oxidation with dichromate standard solution, or similar measurement with potassium permanganate standard solution after alkaline dissolution.[408, 409]

Preparation of the reagent solution. 16 g finely powdered mandelic acid is dissolved in 100 ml water and the solution filtered before use.

Precipitation of the zirconium mandelate complex. The sample solution containing 2–20 mg zirconium per 100 ml is acidified with 50 ml concentrated hydrochloric acid and heated to 95°C. 20–80 ml mandelic acid reagent solution (depending upon the zirconium content of the sample) is added slowly with stirring to the hot sample solution. The mixture is heated for one hour on the water bath and kept for a day. The precipitate is collected on a G4 sintered glass filter, washed with 3×10 ml ethanol, 2×10 ml diethyl ether and dried at 110–120°C. The product is ready for weighing.

Volumetric determination of zirconium mandelate. The complex precipitated and washed as described in the preceding paragraph is dissolved in warm 5 N sodium carbonate solution (4×5 ml sodium carbonate solution is sufficient to dissolve a precipitate containing not more than 20 mg zirconium). The solution is washed with 10–15 ml distilled water into a titration flask, 10 ml 23% sodium hydroxide and 20 ml 0.5 N potassium permanganate solutions are added, and the whole kept for 10 min. The solution is then diluted with distilled water to 300 ml, acidified with 25 ml 50% sulfuric acid, treated with 20 ml 0.5 N oxalic acid solution, heated to about 60°C, and the excess oxalic acid titrated with 0.1 N potassium permanganate standard solution. The same procedure is also carried out without zirconium (blank). The difference between the permanganate consumptions is proportional to the zirconium content to be determined. 1 ml 0.1 N potassium permanganate standard solution corresponds to 0.5136 mg zirconium.[408]

8. FLAVONES

Morin

(3,5,7,2′,4′-Pentahydroxyflavone)

Molecular weight: 238.26
Pale yellow crystalline powder. Melting point 285°C.
Solubility: almost insoluble in water, freely soluble in ethanol and acetone, also soluble in alkaline aqueous media.

The reagent reacts with many metal ions in acidic media to give complexes which are extractable by butanol. Thus, for instance, it is useful for the extraction of aluminium, beryllium, cerium(III), gallium, indium, antimony(III), scandium, tin(IV), thorium, titanium and zirconium.[36]

The fluorescence of the aluminium(III) complex of morin renders possible the quantitative determination of 0.1−1.2 γ aluminium. Gallium, indium, beryllium and rare earth elements interfere with the determination, as the morin complexes of these elements also fluoresce. The interference of the lead, zinc and molybdenum complexes can be eliminated by carrying out the measurement in acetic acid solution as these complexes, due to their lower stabilities, do not form at pH 3. Of the anions, fluoride, which gives an extremely stable complex with aluminium, interferes with the determination; on addition of fluoride, aluminium fluoride is formed and the morin is liberated. The process is accompanied by quenching of the fluorescence. Okáč[328] utilized the phenomenon for the indication of the end-point of the titration of aluminium with fluoride.

Schneer et al.[407, 411] have successfully applied the morin complex of zirconium (absorption max. at 420 nm) for the determination of zirconium on the micro scale.

Morin gives a red precipitate with niobium in acidic solutions, which is useful for the gravimetric determination of niobium. Titanium, zirconium, molybdenum and vanadium interfere with the determination. The tantalum complex of morin also precipitates, but the niobium complex can be separated by extraction with acetone.[371, 456] Both niobium and tantalum can be determined photometrically in the form of their morin complexes.[21]

Morin plays an important role in the analysis of the actinides too.[249, 463]

Morin is used as a universal visualizing agent in paper chromatography.[151, 406]

Quercetin

(3,5,7,3′,4′-Pentahydroxyflavone)

Molecular weight: 338.2
Yellow needles, melting point 316–317°C.
Solubility: practically insoluble in water, sparingly soluble in ethanol and methanol (e.g. 0.4% in abs. ethanol), moderately soluble in hot ethanol (4% in hot abs. ethanol). Soluble in alkaline aqueous medium.

The reagent gives coloured precipitates with chromium(III), aluminium(III), tin(IV) and iron(III). It also reacts with germanium to give a vivid yellow water-soluble complex. Niobium and tantalum react with quercetin in 20–25% sulfuric acid to give red and orange precipitates, respectively. Titanium, which also gives a red precipitate under these conditions, interferes with the determination of niobium and tantalum. The reagent has also been applied for the determination of thorium and molybdenum.[452]

9. β-DIKETONES

Acetylacetone

(Diacetylmethane, 2,4-pentanedione)

$$CH_3-\underset{\underset{O}{\|}}{C}-CH_2-\underset{\underset{O}{\|}}{C}-CH_3$$

Molecular weight: 110.11
Colourless, mobile liquid. Boiling point 135–137°C (at 745 mmHg). Specific gravity: 0.976 at 25°C.
Solubility: 17.2% in water at 20°C. Readily soluble in chloroform, benzene, ethanol and diethyl ether. The distribution ratio in organic solvent–water systems is 3.3 for carbon tetrachloride, 5.8 for benzene and 25 for chloroform.

In aqueous solution acetylacetone acts as a weak monobasic acid, $-\log K_H = 8.9$.

Like other β-diketones, acetylacetone contains an enolic hydroxyl group, whose hydrogen atom is replaceable by a metal, and a ketonic oxygen atom in β-positions to one another. The metal bonds to both oxygen atoms to give a chelate ring.

In addition to acetylacetone, benzoylacetone, dibenzoylmethane and thenoyltrifluoroacetone are the analytically significant representatives of β-diketones. Acetylacetone reacts with almost 60 metal ions to give chelate complexes of stoichiometric composition. These are usually readily soluble in organic solvents, while many of them are volatile; they are relatively stable at higher temperatures and thus can be distilled without decomposition.

Solutions of acetylacetone in carbon tetrachloride, chloroform, benzene or toluene are usually used for the extraction of metal ions, but acetylacetone may also be used without solvent, when it acts simultaneously as solvent and extracting agent. In the latter case the great excess of the reagent necessarily applied renders possible the extraction of metal ions from acidic media where extraction cannot be achieved with solutions of the reagent. In the case of a smaller excess of the reagent, a higher pH value is required in order to ensure the necessary concentration of the free ligand. It is usually advisable, however, to carry out the extraction with a carbon tetrachloride solution of the reagent. In this case the organic phase is the heavier, and separation of the two phases involves less difficulty (no emulsion formation).

In general, the chelates of acetylacetone are formed extremely rapidly. According to earlier investigations, cobalt(II) and nickel(II) do not form extractable chelates with acetylacetone at all. However, the formation of these chelates as well as those of molybdenum(VI) and magnesium(II) requires a fair period of time. The chromium(III) complex forms only on boiling and this makes possible the separation of chromium from almost any other metal ion.

The acetylacetone chelates of iron(III), uranium(VI), vanadium(III), cobalt(III) and chromium(III) are coloured and hence useful for the direct spectrophotometric determination of these metals. The beryllium chelate has its absorption maximum at 295 nm, where acetylacetone itself absorbs also. Excess reagent must therefore be removed before spectrophotometric measurement, or the extraction must be made from some suitable medium which retains the excess reagent.

Starý[443] found that the extractabilities of metal acetylacetone chelates decrease in parallel with their stability constants in the following order: palladium, thallium(III), iron(III), plutonium(IV), beryllium, uranium(IV), gallium, copper(II), scandium, aluminium, indium, uranium(VI), thorium, lead, nickel, lanthanum, cobalt(II), zinc, manganese and magnesium.

The metal ions extractable with acetylacetone and the optimum extraction conditions (pH, solvent) are summarized in Table 21. These data illustrate the widespread application of acetylacetone in analytical chemistry.

Purification of the reagent. Commercial acetylacetone usually contains 2–15% acetic acid. This used to be removed by washing with dilute ammonium hydroxide solution. Ammonia was then removed from the reagent by washing with a little water, followed by drying over anhydrous sodium sulfate and distillation.[296] This purification method supplied pure acetylacetone, which could be stored for long periods without decomposition, but the purification was accompanied by relatively great losses of the reagent. Rydberg[385] therefore suggested the

TABLE 21. METAL IONS EXTRACTABLE WITH ACETYLACETONE

Metal ion	pH of the aqueous solution		Absorption maximum, nm	Refs.
	extracted with acetylacetone	extracted with a mixture of an organic solvent † and acetylacetone		
Al(III)	3–6	5–9 (benzene)		443
Be(III)	1.5–3	3.5–8.0 (benzene)	295 ($\varepsilon = 31{,}600$)	343, 443, 446
Cr(III)	formed at boiling temperature		560	312
Cu(II)	2–5	4–10 (benzene)		257, 443, 446
Fe(III)	1	2.5–7.0 (benzene)	440	257, 443, 444, 447
Ga(III)	6–9	3.5–8 (benzene)		443
Hf(IV)	> 3	≈ 7 ($CHCl_3$)		386
In(III)	3–6	> 5.5 (CCl_4)		383, 447
Mo(VI)	6–0.01 M H_2SO_4		352 ($\varepsilon = 1630$)	292
Nb(V)		2–5 (benzene)		451
Pb(II)	6–8	7–10 (benzene)		256, 443
Pd(II)		0–8 (benzene)		443
Pu(IV)		4–7 (benzene or $CHCl_3$)		386
Ru(III)		4–6 (heating) (benzene)	505	75
Sc(III)		3.5–9 (benzene)		443
Th(IV)		5–9 (benzene)		342, 443
Tl(III)		2.0–10 (benzene)		443
U(IV)		> 3		386, 387
U(VI)	2–7			256
V(III)		2.3–3.0 (A : $CHCl_3$ = 1 : 1)	390	293
V(IV)		2–4 (A : $CHCl_3$ = 1 : 1)		293
V(V)		2.1 (A : $CHCl_3$ = 1 : 1)	355 450	293
Y(III)	5.5–10.0			428
Yb(III)	∼ 6			79
Zn(II)	5.5–8			257, 446
Zr(IV)	∼ 2	3–8 ($CHCl_3$)		257, 293

† The diluting solvent is shown in parentheses

following procedure. 20 ml crude acetylacetone is dissolved in 80 ml benzene and the solution extracted three times with 100 ml distilled water. Acetic acid contamination passes quantitatively into the aqueous phase during the three extractions, while acetylacetone remains in the benzene phase. The resulting benzene solution of acetylacetone can be used directly in most cases, while if necessary benzene can be removed by distillation.

Determination of beryllium. The sample solution containing about 10 γ beryllium per 50 ml is acidified to about pH 1, 2 ml 10% EDTA solution is added and the mixture neutralized with 0.1 N sodium hydroxide to pH 7–8. After addition of 5 ml 5% aqueous acetylacetone solution, the pH of the mixture is adjusted to 7–8, the solution kept for 5 minutes and extracted three times with 10 ml chloroform. The extinction of the chloroform solution is measured at

295 nm, the absorption maximum of the beryllium acetylacetone chelate. By this method beryllium can be determined in the presence of thousand-fold amounts of iron, aluminium, chromium, zinc, copper, manganese, lead, silver, cerium and uranium.[2]

Determination of chromium. The sample solution containing chromium(III), with pH about 3–4, is extracted with a 1:1 mixture of acetylacetone and chloroform to remove accompanying metal ions. The pH of the aqueous phase is adjusted to about 6, 10 ml acetylacetone is added to about 50 ml sample solution and the mixture is refluxed for 1 hour. During this time the chromium complex is quantitatively formed. The cooled solution is acidified with strong mineral acid (to 1–3 N for hydrogen ion). The chromium complex can now be extracted from the solution with a mixture of acetylacetone and chloroform. The absorption maximum of the complex is at 60 nm.[312]

Determination of cobalt. The aqueous sample solution containing cobalt(II), with pH 6–7, is extracted with a 1:1 mixture of acetylacetone and chloroform to remove accompanying metals. The aqueous phase is then treated with a few millilitres of acetylacetone and 5 ml 3% hydrogen peroxide. The hydrogen peroxide oxidizes the cobalt(II) to the cobalt(III) complex of acetylacetone. For quantitative transformation, the solution is basified to pH 8–9 and refluxed for 10 minutes. After cooling, the solution is acidified to about pH 1 and the cobalt(III) complex is extracted with a 1:1 mixture of acetylacetone and chloroform. The complex is a vivid green colour.[312]

Determination of iron. The iron(III) complex forms quantitatively even at pH 1, and it can also be extracted quantitatively with a 1:1 mixture of acetylacetone and chloroform. The absorption maximum of the complex is at 440 nm. By this method iron can be determined in the presence of many other metal ions, but beryllium, molybdenum, tungsten, copper and gallium interfere with the determination.[292]

The thermal stability and relatively high volatility of the metal complexes of β-diketones[431] have rendered possible the introduction of the gas-chromatographic separation of these chelate complexes in analytical practice.[313] Experience has revealed the particularly advantageous volatility features of the metal complexes formed with fluorine-substituted diketones. Thus, for example, trifluoroacetylacetone has been applied with outstanding results.[440]

Belcher,[41] for instance, successfully separated the *cis* and *trans* isomers of the trifluoroacetylacetone chelate of chromium. The method has also proved suitable for the gas-chromatographic separation of aluminium, chromium and iron.[44] Trifluoroacetyl-pivaloyl-methane (1,1,1,-trifluoro-5,5-dimethyl-2,4-hexanedione) has proved to be a most suitable chelating agent for the gas-chromatographic separation of metals.

Lippard[270, 271] and later Belcher *et al.*[45] showed that 1:3 metal:ligand complexes of rare earth metals with diketone ligands substituted by fluorine to various extents, react with alkali metal ions to give complexes of composition $M'ML_4$, where M' is the alkali metal ion, M the rare earth metal ion and L the diketone ligand. The gas-chromatographic separation of rare earth metals in the form of these complexes could be achieved.

It should be noted at this point that gas-chromatographic separation of chelate complexes appears to be a challenging area of analytical chemistry, particularly for the future; however, elaboration of accurate, routine-like analytical procedures requires further thorough investigations in this field.

Thenoyltrifluoroacetone (TTA)

$$\underset{S}{\bigcirc}-\underset{\parallel}{\overset{}{C}}-CH_2-\underset{\parallel}{\overset{}{C}}-CF_3$$
$$\quad\quad\; O \quad\quad\; O$$

Molecular weight: 222.2
Pale yellow, crystalline compound. Melting point 42.5–43.2°C.
Solubility: sparingly soluble in water, readily soluble in organic solvents.
Distribution ratio between acidic water and benzene is about 40.
In alkaline aqueous medium it exists in the enolate form, this being relatively more soluble in water. The equilibrium distribution ratio between benzene and alkaline aqueous medium (about pH 8) is about 1. In alkaline aqueous medium of pH >9 TTA decomposes to trifluoroacetone and acetylthiophen.[128]

Thenoyltrifluoroacetone is one of the most proved extracting agents in solvent-extraction analysis. It is equally suitable for the separation and selective enrichment of metal ions. Table 22 summarizes the analytical applications of the reagent. One of the reasons for its widespread analytical applicability is that due to the electron-attracting effect of the trifluoromethyl group the acidity of the enolate form is much more pronounced than those of related reagents containing no fluorine atoms (pK_H of TTA is 6.23). This is the reason why metal ions can be extracted with TTA from more acidic solutions than with other β-diketones. The phenomenon also makes possible the extraction of metal ions which undergo hydrolysis in acidic media.[374] Thenoyltrifluoroacetone has proved to be particularly advantageous in the separation and enrichment of actinides.

Complexes of TTA usually have the composition $M(TTA)_n$, where n is the charge of the metal ion, but there are two exceptions: the strontium(II) and uranium(VI) complexes have compositions $M(TTA)_n HTTA$.[252, 344]

In solvent extraction procedures 0.1–0.5 M solutions of TTA in benzene or toluene are usually used, but in some cases xylene or methyl isobutyl ketone are also employed as solvents.

Some TTA complexes absorb in the visible region and accordingly are coloured. Thus the uranium(VI) complex is yellow, the copper(II) complex green, the iron(III) complex red, the chromium(III) complex deep yellow and the cerium(IV) complex deep red. In these cases the reagent is also suitable for the photometric determination of these metals. Thenoyltrifluoroacetone itself has an absorption maximum at 330 nm with a molar extinction coefficient of 11,900.

Some of the more important solvent extraction procedures are given below.

Extraction of actinium. The aqueous acidic solution (pH 5.5) containing actinium among other metals is extracted with a 0.25 M solution of TTA in benzene. The organic phase is then extracted with dilute nitric acid (pH 4) to separate actinium from the metal ions which accompanied actinium in the first extraction procedure. These metals (e.g. lead, bismuth, thorium, etc.) remain in the organic phase during the second extraction procedure, while actinium passes back to the aqueous phase. The pH of the aqueous phase is then increased to 5.5 and the solution is extracted again with 0.25 M thenoyltrifluoroacetone in benzene.[199]

Extraction of rhodium. 2 ml of a sample solution containing not more than 20 mg rhodium(III) is treated dropwise with 10 M sodium hydroxide solution until the appearance of a precipitate; this is then dissolved by addition of one drop of concentrated perchloric acid. 10 ml 0.5 M TTA solution (in acetone) is added to the mixture, followed by 2 ml acetic–acetate

TABLE 22. OPTIMUM CONDITIONS FOR THE EXTRACTION OF
THENOYLTRIFLUOROACETONE COMPLEXES

Central atom	pH of the aqueous solution	Organic solvent	HTTA concentration, M	Extraction constant, log K	Refs.
Ac(III)	> 5.5	benzene	0.25		199
Al(III)	5.5–6.0	4-methyl-2-pentanone	0.10		153
	5.5–6.0	benzene	0.1	− 5.23	72
Ba(II)	∼ 8	benzene	0.5	−14.4	357
Be(II)	∼ 4	xylene	0.5–1.0		136
Bi(III)	> 2.5	benzene	0.25		199
Ca(II)	∼ 8	benzene	0.1	−12.0	72
Cd(II)	∼ 7	$CHCl_3$	0.1	−11.4	423
Ce(III)		benzene		− 9.43	357
Cf(III)		toluene	0.2	− 7.8	277
Cm(III)		toluene	0.2	− 8.6	277
Co(II)	7.6–8.8	isobutyl alcohol	0.1		493
Cr(III)	5–7	benzene	0.15		279
Cu(II)	3–6	benzene	0.15	− 1.32	244
Dy(III)	∼ 3	benzene	0.5	− 7.03	218
Eu(III)	∼ 3	benzene	0.5	− 7.66	357
Fe(III)	∼ 2	benzene	0.15	3.3	357
Fm(III)	3.4	toluene	0.2	− 7.7	277
Gd(III)	> 3	benzene	0.5	− 7.58	357
Hf(IV)	2 M $HClO_4$	benzene		7.8	212
Ho(III)	> 3	benzene	0.2	− 7.25	357
In(III)	2.5–3.5	benzene	0.5	− 4.34	357
Ir(III)	> 7	benzene			357
La(III)	> 3.5	benzene	0.5	−10.51	357
Lu(III)		benzene		− 6.77	357
Nb(V)	10 M HNO_3	xylene	0.5		306
Nd(III)	> 3	benzene	0.5	− 8.57	357
Ni(II)	5.5–8.0	benzene : acetone (1:3)	0.15		135
Np(IV)	1 M HCl	xylene	0.5	5.6	305
Np(V)	7–9	isobutyl alcohol	0.01		493
Pa(V)	2–6 M HCl	benzene	0.5		318, 357
Pm(III)		benzene	0.5	− 8.05	357
Pr(III)	> 4	benzene	0.1–0.5	− 8.48	243
Pu(III)		benzene		− 4.44	357
Pu(IV)	0.5–1.0 M HNO_3	benzene	0.5–1.0		247
Pu(VI)		benzene		− 1.82	357
Sc(III)	> 1.6	benzene	0.5	− 0.77	357
Sm(III)	∼ 3–4	benzene	0.5	− 7.68	357
Sr(II)	10–12	methyl isobutyl ketone	0.2		252
Tb(III)		benzene		− 7.51	357
Th(IV)	> 1	benzene	0.25–0.45	− 0.8	199
Tl(I)	∼ 7		0.25	− 5.2	199
Tl(III)	∼ 4	benzene	0.25		199
Tm(III)		benzene	0.5	− 6.96	357
U(IV)	1 M $HClO_4$	benzene		5.3	55

TABLE 22 (continued)

Central atom	pH of the aqueous solution	Organic solvent	HTTA concentration, M	Extraction constant, log K	Refs.
U(VI)	3.5–8	benzene	0.15		245
V(IV)	~ 4	benzene	0.25		425
Y(III)	6–9	benzene	0.10	− 7.39	357, 425
Yb(III)		benzene		− 6.72	357
Zr(IV)	2 M HClO$_4$	benzene	0.5	9.0–9.15	127

buffer solution (1 ml M acetic acid–1 ml M sodium acetate). The solution is heated for about 20 min on the water bath in order to remove acetone, transferred quantitatively to a separating funnel and extracted with 30 ml benzene. In order to remove accompanying metal ions, the organic phase is washed successively with 15 ml 6 M hydrochloric acid, 15 ml M sodium hydroxide, 15 ml 8 M nitric acid, and finally 15 ml 6 M hydrochloric acid solutions.[269]

Extraction of thallium(III). Thallium(III) can be extracted with thenoyltrifluoroacetone solution in benzene from aqueous solutions of pH > 3.8. Under such conditions, the TTA complex of bismuth(III) also enters the organic phase. The thallium(III) is therefore re-extracted from the benzene solution with 0.01 N nitric acid, when the bismuth remains in the benzene phase. After adjustment of the pH of the aqueous phase to >3.8 thallium(III) is extracted with TTA as above.

The interfering effect of lead may be eliminated by reduction of thallium(III) to thallium(I). Lead can be extracted at pH 5 with 0.25 M TTA solution in benzene, when the thallium(I) remains in the aqueous phase. The latter is then extracted with thenoyltrifluoroacetone at pH 7–8.[199]

Extraction of zirconium. The outstanding stability of the thenoyltrifluoroacetone complex of zirconium renders possible the extraction of this metal with 0.5 M thenoyltrifluoroacetone solution in xylenol even from 6 M hydrochloric acid or 2 M nitric acid solutions. Anions giving highly stable complexes with zirconium, such as oxalate, fluoride, sulfate and phosphate, interfere with the extraction. Extraction from 6 M hydrochloric acid solution is specific for zirconium even in the presence of aluminium, iron, rare earth metals, thorium and uranium.[304]

10. DITHIZONE

(Diphenylthiocarbazone)

$$S=C\begin{cases}NH-NH-C_6H_5\\N=N-C_6H_5\end{cases}$$

Molecular weight: 256.3
Violent-black crystalline powder. Melting point 165–169°C (with decomposition).
Solubility: sparingly soluble in water and dilute mineral acids, dissolves in dilute ammonia to give a yellow colour (enolization also occurs). Sparingly soluble in hydrocarbons, moderately soluble in chloroform and carbon tetrachloride. For example, 6.8×10^{-2} M and 2.5×10^{-3} M solutions can be prepared with chloroform and carbon tetrachloride.[253] In analytical practice chloroform

and carbon tetrachloride solutions of dithizone are used almost exclusively. The distribution ratio between organic solvents and water are extremely high (carbon tetrachloride: 1.1×10^4, chloroform: 2×10^5.[397]

Dithizone acts as a monobasic acid in aqueous solution. The acidic dissociation constant is 3.2×10^{-5}.[219]

Dithizone may react in either of two forms: structures I and II show the enol and the keto forms, respectively.

$$\text{HS—C}\begin{array}{c}\text{N=N—}\\ \\ \text{N=NH—}\end{array} \qquad \text{S=C}\begin{array}{c}\text{N=N—}\\ \\ \text{NH—NH—}\end{array}$$

$$\text{I} \qquad\qquad\qquad \text{II}$$

Thus, heavy metals give two types of complex. According to the investigations of Irving,[220] S-methylated dithizone does not react with heavy metal ions. This observation proves that the sulfur atom acts as donor in the primary dithizonates (called also keto-dithizonates). X-ray diffraction studies,[80, 266] too, indicate that the structure of primary dithizonates corresponds to the following formula:

$$\left[M \begin{array}{c} \text{S—C—N=N—C}_6\text{H}_5 \\ \| \\ \text{HN—N} \\ | \\ \text{C}_6\text{H}_5 \end{array} \right]_n$$

The structure of complexes formed by the enol tautomer of dithizone are not yet fully elucidated. On the basis of the stoichiometry of the complexes it has been assumed that dithizone acts in these compounds as a diprotic acid. However, the assumption has never been categorically confirmed. In the course of a study of the primary and secondary forms of the copper dithizonate complex, Freiser et al.[177] showed that the so-called "enol form", the secondary dithizonate, is diamagnetic and contains a copper(I) central atom; this is the reason for the 1:1 metal:ligand composition in the complex. Consequently, dithizone acts in this complex as a monoprotic acid. The examinations of Briscoe et al.[78] have led to analogous conclusions. These authors found the presence of a mixed complex containing a chloride ion and a dithizone ligand in the aqueous solution of the mercury secondary dithizonate complex.

Accordingly, elucidation of the actual structures of complexes formed with the enol form of dithizone requires further examinations.

The primary complexes of dithizone possess much greater analytical significance. These are much more stable than the secondary complexes and are also more soluble in organic solvents. From an analytical viewpoint it is particularly significant that primary complexes form in acidic solutions, containing the ligand in excess, while the secondary complexes form in the presence of an excess of the metal ion. (Higher pH favours also the formation of secondary complexes.)

Dithizone is known to give analytically applicable chelate complexes with the following metal ions: manganese(II), iron(II), cobalt(II), nickel(II), copper(I), copper(II), silver(I), gold(III), palladium(II), platinum(II), zinc(II), cadmium(II), mercury(I), mercury(II), gallium(III), indium(III), thallium(I), tin(II), lead(II), bismuth(III), tellurium(IV) and plutonium(IV).

The difference between the stabilities of chelate complexes formed by various metal ions and the different extraction possibilities determined by the former, make possible the separation of several metal ions (see Table 23). Thus, for instance, silver, mercury, copper, palladium and gold can be separated from other metals in dilute mineral acidic media (0.1–0.5 M) with dithizone.

The selectivity of extraction can be further enhanced by application of masking agents. For instance, only zinc(II), tin(II) and palladium(II) can be extracted quantitatively in the presence of excess thiosulfate, and only relatively large amounts of cadmium(II), cobalt(II) and nickel(II) interfere with their determination. In the presence of cyanide as masking agent, only zinc(II) and tin(II) can be extracted. EDTA as masking agent makes possible the separation of silver or mercury from copper. Lead(II), zinc(II), bismuth(III), cadmium(II), nickel(II), cobalt(II) and thallium(I) can be masked with EDTA.[180]

The selectivity of extraction of metal ions by dithizonate is also increased by the fact that some metal complexes are inert. Thus, for example, cobalt(II) and nickel(II) react with dithizone only in alkaline media, but the chelates once formed are resistant to acids. Hence, all those metal dithizonate complexes which are decomposed by acid can be extracted by mineral acid solutions from the cobalt and nickel dithizonates. On the other hand, the dithizonate complexes which form in acidic solutions can be removed before formation of the cobalt and nickel complexes.

Dithizone acts as reducing agent towards oxidizing cations. It has been found that complexes of dithizone with iron, tin and platinum contain the metals in their lower oxidation states, even if the dithizone was treated with iron(III), tin(IV) or platinum(IV) salts.

Purification of the reagent. Dithizone is extremely sensitive towards oxidation, and commercial preparates are therefore usually contaminated by the oxidation product of dithizone, diphenylthiocarbodiazone. The presence of the latter can be checked as follows: a 0.01% solution of dithizone in carbon tetrachloride is extracted with dilute ammonia solution. The organic phase must be colourless or a very pale yellow. If not, purification may be performed in the following manner:

A saturated solution of dithizone in chloroform is concentrated at 40°C until about 50% of the dithizone content crystallizes. The crystalline product is collected on a sintered glass filter, washed with a little carbon tetrachloride and dried by the passage of a stream of air by suction. This manner of purification is fairly simple and reliable, but it is accompanied by a relatively great loss of the material. A more economical procedure for the purification of dithizone is the following: 0.5 g dithizone is dissolved in 50 ml chloroform and the solution is washed four times with 50 ml 0.5% ammonia solution. The aqueous phase is filtered and treated with either hydrochloric acid or sulfur dioxide gas to precipitate dithizone, which is then extracted with 15–20 ml portions of chloroform. The combined chloroform solutions are washed with water and evaporated to dryness at 50°C. The residue is dried in a desiccator and stored in the dark.

It is not always necessary to use pre-purified dithizone; purification can be performed in one step in the preparation of the dithizone reagent solution. Commercial dithizone is dissolved in chloroform and the solution is extracted with ammonia as above; the decomposition products remain in the organic phase, while dithizone passes into the aqueous phase. The latter is filtered, neutralized with 10% hydrochloric acid and the precipitate extracted with chloroform. The concentration of the dithizone reagent solution thus obtained is simply determined spectrophotometrically: the extinction of the solution measured at 606 nm in a 1 cm cell is divided by 40.6×10^3, the molar extinction coefficient.

Determination of bismuth. 0.5–1.0 g of metal sample containing about 0.01% bismuth is dissolved by appropriate treatment with nitric acid. The nitrogen oxides are removed by boiling, the solution is concentrated to about 25 ml, 1–2 g ammonium citrate is dissolved in it, and the solution is neutralized with ammonia. After dissolution of 2 g potassium cyanide as masking agent in the solution, it is treated with 5 ml 10% ammonia and extracted four times with 15 ml 0.1% dithizone solution in chloroform. The absorption maximum of the complex is at 500 nm in chloroform solution.[198]

Determination of cobalt. 10 ml sample solution containing 1–50 γ cobalt is treated successively with 5 ml 10% sodium citrate solution and concentrated ammonium hydroxide to give pH 8. The resulting solution is vigorously shaken with 10 ml 0.01% dithizone solution in carbon tetrachloride for about half a minute and the two phases are separated. In order to ensure the quantitative removal of the cobalt dithizonate complex, the aqueous phase is extracted with a further 2–3 ml dithizone reagent solution, and the procedure is repeated until the carbon tetrachloride phase remains colourless after shaking for 1 minute.

The procedure permits the separation of cobalt from iron(III), titanium(III), chromium(III) and vanadium salts. Inasmuch as the cobalt sample contains copper(II) or other metal ions which form dithizonate complexes in acidic medium, these metals are removed by means of a 0.2% solution of dithizone in chloroform at pH 3–4 (citrate), prior to formation of the cobalt dithizonate complex.[285, 399]

Determination of lead. The slightly acidic sample solution containing about 100 γ lead in 25 ml is treated with 75 ml of the following reagent solution: 350 ml concentrated ammonium hydroxide, 30 ml 10% potassium cyanide solution and 1.5 g sodium sulfite in 1 litre total volume.

2.5 ml 0.005% dithizone solution in chloroform and 17.5 ml chloroform are added to the above 100 ml aqueous solutions, the mixture is thoroughly shaken for 1 minute and the two phases are separated. The absorption maximum of the lead dithizonate complex appears at 510 nm in chloroform solution. Only bismuth and thallium interfere with this determination of lead. As concerns anions, the presence of more than 5 mg phosphate also causes error.[436]

Determination of zinc. The acidic sample solution containing about 5 γ zinc in 10–25 ml is treated with 0.5 M sodium acetate solution until the pH of the solution is 5.0–5.5. The interfering effects of mercury(II), copper(II), bismuth(III), silver(I) and lead(II) can be eliminated by sodium thiosulfate. In masking these accompanying ions, 750 mg crystalline sodium thiosulfate is added for each mg mercury content, 600 mg per mg copper, 350 mg per mg bismuth, 60 mg per mg silver and 40 mg per mg lead. The aqueous phase is then extracted repeatedly with 3 ml 0.05% dithizone solution in carbon tetrachloride until the colour of the last dithizone solution extract remains unchanged during extraction for three minutes. The combined organic phase is washed three times with 5 ml of a solution containing 225 ml 0.5 M sodium acetate, 50 ml 10% sodium thiosulfate and 40 ml 10% nitric acid in 500 ml total volume, once with distilled water and then with 5 ml 0.04% sodium sulfide solution until the sodium sulfide washings remain colourless. The absorption maximum of the zinc complex remaining in the organic phase is at 535 nm.

Interferences from nickel and cobalt can be eliminated by cyanide as masking agent. The cyanide solution is added to the reaction mixture before adjustment of the pH with sodium acetate. 15 ml 5% sodium cyanide solution is sufficient to mask about 100 mg nickel and cobalt.

In the presence of aluminium, zinc cannot be extracted quantitatively from the above-specified slightly acidic solution. Therefore, in the presence of more than 100 mg aluminium

TABLE 23. APPLICATION OF DITHIZONE FOR DETERMINATION OF METAL IONS BY SPECTROPHOTOMETRY COMBINED WITH SOLVENT EXTRACTION

Metal ion	pH of the aqueous phase	Extracting solvent	Absorption maximum	Molar extinction coefficient	Extraction constant,† $\log K$	Interfering ions	Successful application	Note, masking agent	Refs.
Ag(I)	4 M H_2SO_4, 7	CCl_4	426	30,500	7.16	only Pd, Au, Hg, Cl^-, Br^-, I^-, CN^-	analysis of metals, ores and alloys		131, 166, 224, 242, 253, 434
Au(III)	0.5 M H_2SO_4	$CHCl_3$	450	24,000		only Ag, Hg, Pd, much Cu	analysis of ores and alloys		150, 163, 221, 434
Bi(III)	3–10	CCl_4	490	80,000	9.98	Pb, Te(I), Sn(II)	determination of Bi in high--purity substances	CN^-	103, 211, 229, 230
Cd(II)	6.5–14	CCl_4	520	88,000	2.14	high concn. of metal ions	measurement of Cd traces after separation		20, 165, 230, 317, 441
Co(II)	5.5–8.5	CCl_4	542	59,200	1.6	high concn. of metal ions	examination of biological substances		230, 253, 412
Cu(II)	1–4	CCl_4	550	45,200	10.53	only Hg, Ag, Au, Pd, much Bi	analysis of Fe, steel, high-purity Al, Ni, uranium ore, and biological substances		146, 164, 186, 230, 311, 369
Fe(II)	7–9	CCl_4	520			no analytical importance			133, 230

Hg(II)	6 M H_2SO_4, 4	CCl_4	485		26.8	only Ag, Pd, Au, Pt, much Cu		1, 77, 145, 299
In(III)	5–6.3	CCl_4	510	71,200	4.84	Bi, Pb, Sn(II), Tl(I), Hg, Ag, Pd, Au, Cu	analysis of uranium and thorium alloys	CN^- 1%	16, 289, 346
Ni(II)	6–9	CCl_4	665	87,000	−0.6	high concn. of metal ions	analysis of uranium alloys and silicates		230, 310, 347, 441
Pb(II)	8.0–10	CCl_4	520	19,200	0.44	Bi, Tl(I), Sn(II)	trace analysis in alloys, drugs, and organic substances	CN^- tartrate, citrate	23, 125, 253, 326
Pd(II)	0	CCl_4	620	68,800		Ag, Hg, Au			230, 489
Pt(II)	1–10.5 N H_2SO_4	benzene	720				determination of alloying platinum in Au		241, 300
Tl(I)	11–14.5	$CHCl_3$	505	27,000	−3.5	Pb, Bi, Sn(II), Th	analysis of alloys	CN^-	230, 346, 432
Zn(II)	6–9.5	CCl_4	535	33,600	2.0–2.3	high concn. of metal ions	analysis of metals, alloys and biological substances after separation and masking	$S_2O_3^{2-}$ masks Cu, Hg, Ag, Au, Bi, Pb	31, 129, 221

Fe(III) and Sn(IV) do not give extractable complexes with dithizone.

† $K = \dfrac{[MA_N]_{org}[H]^N}{[M][HA]_{org}^N}$, where $[MA_N]_{org}$ = metal dithizonate concentration in the organic phase;

$[HA]_{org}$ = dithizone concentration in the organic phase.

the zinc content is extracted from slightly ammoniacal medium containing citrate. Under such conditions, however, other heavy metals too pass into the organic phase together with zinc. Nevertheless, the relatively low stability of the zinc dithizonate complex makes possible the separation of the zinc from the accompanying heavy metal ions by extraction with 0.1 N hydrochloric acid, when only the zinc passes back to the aqueous phase. 0.1 N hydrochloric acid decomposes the zinc dithizonate complex, while the corresponding dithizonate complexes of the accompanying heavy metals do not decompose or do so much more slowly. The separated zinc can now be determined as described above.

Determination of nickel. The pH of the sample solution containing about 15 γ nickel in 10 ml is adjusted to 10 with an ammonium chloride–ammonium hydroxide buffer solution. The aqueous solution is extracted with 15 ml of a solution of dithizone and phenanthroline (6×10^{-4} and 3×10^{-4} M, respectively) in chloroform by shaking thoroughly for 2 minutes. After separation of the two phases the chloroform solution is washed with 20 ml 0.2 M sodium hydroxide solution, filtered through a dry filter paper and its absorption measured at 520 nm in a 1 cm cell. The above dithizone–phenanthroline solution in chloroform is used as reference solution.[175]

On the basis of the data summarized in Table 23 one can easily prepare analogous procedures for the determination of several other metal ions as well. The Table gives all the necessary information concerning the optimum conditions of extraction (pH, solvent, and in some cases the masking agent), the data necessary for the spectrophotometric measurements (the wavelength of the absorption maximum and the value of the molar extinction coefficient), and also the extraction constants and disturbing ions, i.e. the selectivity of the extraction of an individual metal ion. Hence the Table can be used to perform separations which are not included above.

Busev *et al.*[104] have introduced a novel analytical application of dithizone. They achieved the separation of several metal ions by chromatography of their dithizone complexes in non-aqueous medium.

11. DITHIOCARBAMATES

Sodium diethyldithiocarbamate

$$\begin{array}{c} H_5C_2 \\ \diagdown \\ N-C \\ \diagup\diagdown \\ H_5C_2 SNa \end{array}$$

Molecular weight: 171.25
White crystalline compound.
Solubility: about 35% in water, less soluble in organic solvents. However, the free acid, diethyldithiocarbamic acid, is readily soluble in organic solvents and less soluble in water. Thus, on acidification of an aqueous solution of a diethyldithiocarbamate, the free acid can be extracted with chloroform or carbon tetrachloride. The distribution ratio is 2,360 for chloroform and 343 for carbon tetrachloride.[71]

The $-\log K_H$ of the acid is 3.4 in aqueous medium. From this fact and the distribution ratio it follows that the reagent can be extracted quantitatively from acidic media (pH < 4) with organic solvents, while at pH > 8 it remains quantitatively in the aqueous phase.

Sodium diethyldithiocarbamate reacts with even more metal ions than does dithizone. However, its analytical application is seriously limited by its considerably lower stability in acidic aqueous media. According to the examinations of Bode,[71] the reagent undergoes considerable decomposition within 5 minutes at pH 5. The analytical application of the reagent is therefore restricted to a very narrow pH range. There is no chance with this reagent to enhance the analytical selectivity of extractions by making use of the difference between the stabilities of its complexes with various metal ions and to choose the pH of the reaction mixture accordingly. Notwithstanding, some authors suggest extraction procedures from acidic media, for instance in the cases of such stable complexes as those of copper(II), nickel(II) and bismuth(III). However, in these procedures extraction must be done *immediately* after acidification of the solution, and even then the error caused by decomposition of the reagent must be taken into account.[264] The analytical selectivity of the complexation reactions of diethyldithiocarbamate can be enhanced by using various auxiliary complexing and masking agents. EDTA is particularly useful for this purpose, but there are also other complexing agents, such as cyanide or citrate.

The reagent does not absorb in the visible region (>400 nm). On the other hand, its metal complexes are generally coloured, which makes possible the application of the reagent for spectrophotometric measurements. For instance, its cobalt complex is green (absorption maximum at 650 nm), the copper complex brown (absorption maximum at 440 nm), the iron(III) complex red (515 nm), etc.

The data in Table 24 reveal the specific conditions under which various metal ions can be extracted in the form of their complexes with diethyldithiocarbamate. The Table also contains information concerning the selectivity of extraction of the various metal ions. It can be seen that diethyldithiocarbamate compares favourably with dithizone and oxine, as one of the most useful extracting and spectrophotometric agents in analytical chemistry. In the following some analytical procedures using diethyldithiocarbamate are described. From the data in Table 24, the practical analyst can successfully accomplish further analyses which are not described here. Besides being well proved in practice, the procedures presented may also serve as illustrations of how to develop other solvent extraction analytical procedures from the data in the Table.

Determination of antimony. 10 ml 5% aqueous EDTA solution is added to the sample solution containing not more than 300 γ antimony(III) in 10 ml, and the pH of the solution is adjusted to 9. After addition of 5 ml 10% sodium cyanide solution, the pH of the mixture is checked and adjusted to 9.0–9.5 if necessary. 1 ml 0.2% sodium diethyldithiocarbamate solution and then 10 ml carbon tetrachloride are added, and the mixture is thoroughly shaken for 1 minute. The absorption maximum of the antimony complex is at 350 nm in carbon tetrachloride solution. Bismuth(III), thallium(III) and large amounts of mercury(II), arsenic(III) and copper(II) interfere with the determination.[69]

Determination of bismuth. The selective determination of bismuth with this reagent is performed in the presence of EDTA and sodium cyanide by extraction from dilute ammoniacal solution. The masking solution is prepared by dissolving 50 g EDTA and 50 g sodium cyanide in 1 litre 1.5 M ammonium hydroxide. Depending upon the concentration of the accompanying ions, 10–20 ml of the masking solution is added to the 10 ml sample solution containing 50–300 γ bismuth, followed by 1 ml 0.2% aqueous sodium diethyldithiocarbamate solution and 10 ml carbon tetrachloride. The mixture is vigorously shaken for about half a minute and the absorption of the complex in the carbon tetrachloride solution is measured at 400 nm.[118]

TABLE 24. APPLICATION OF DIETHYLDITHIOCARBAMATE FOR DETERMINATION OF METAL IONS BY SPECTROPHOTOMETRY COMBINED WITH SOLVENT EXTRACTION

Metal ion	pH of the aqueous solution	Extracting solvent	Extinction maximum, nm	Molar extinction coefficient	Interfering ions	Successful application	Notes, masking agent	Refs.
Bi(III)	4–11	CCl_4	366–370	8,620	Tl(III) only in the presence of EDTA, at pH > 11	analysis of alloys		67, 69, 118, 323
Co(III)	4–11	CCl_4	367 650	15,700 549	as the Co complex is inert, the other metal ions can be separated by acid extraction	determination of Co in Ni, steel and rocks	Bi is masked by EDTA + KCN in solutions of pH > 8	69, 362, 363, 448
Cu(II)	4–11	CCl_4	436	13,000	Bi, Tl(III), Au(III), much Pd and Pt, Os, CN^-	analysis of Ni and Co salts, alloys, soil and biological substances		69, 170, 239, 246, 298, 355
Mn(II)	6–9	CCl_4	355 505	9,520 3,710	high concn. of metal ions, EDTA	analysis of steel	many interfering ions can be masked with KCN at pH 7–9	67, 137, 297
Nb(V)	4–5	CCl_4			F^-, citrate	determination of Nb in the presence of Ti, Zr, V and Be		69, 191

Ni(II)	5–11	CCl$_4$ isoamyl alcohol	326 325	34,200 37,000	EDTA, CN$^-$		67, 126
Sb(III)	4–9.5	CCl$_4$	350	3,370	in the presence of EDTA and CN$^-$, at pH 8–9.5, only Bi, Te and Tl interfere		67, 280
Sn(IV)	4–6.2	CCl$_4$ or CHCl$_3$	300–500			EDTA, citrate and tartrate are suitable to mask interfering ions	280
Te(IV)	4–8.8	CCl$_4$	428	3,160	in the presence of EDTA and CN$^-$, at pH 8.5–8.8, only Bi, Sb(III), and Tl(III) interfere	separation of Te(IV) from Te(VI)	Te cannot be extracted at pH > 10	67, 68, 69
Tl(III)	4–11	CCl$_4$ or CHCl$_3$	426	1,330			CN$^-$ is suitable to mask interfering ions	67, 69
V(V)	3–6	CCl$_4$	400	3,790			unextractable at pH > 7	67, 69

Ba(II), Ca(II), Cr(III), La(III), Mg(II), Se(III), Sr(III), Ta(V), Ti(IV), Th(IV), Y(III) and Zr(IV) cannot be extracted with diethyldithiocarbamate. The diethyldithiocarbamates of other metals not listed in the Table cannot be determined directly by spectrophotometry after extraction.

Determination of palladium. The aqueous sample solution (10 ml) containing 200–300 γ palladium is treated with 10 ml 5% EDTA solution, the pH of the mixture is adjusted to 11, 1 ml 0.2% sodium diethyldithiocarbamate reagent solution is added and the mixture is extracted with 10 ml carbon tetrachloride by vigorous shaking for about 1 minute. The absorption maximum of the palladium complex is at 305 nm in carbon tetrachloride solution. Copper(II), silver, mercury(II), bismuth and thallium(III) interfere with the determination of palladium in this manner, as all these ions are extractable from the aqueous phase together with palladium under the conditions specified.[69]

Determination of thallium(III). The sample solution containing not more than 300 γ thallium(III) in about 10 ml is treated with 10 ml 2 M ammonium hydroxide solution containing 5% EDTA and 5% sodium cyanide. The pH of the mixture is adjusted to 11, 1 ml 0.2% aqueous sodium diethyldithiocarbamate solution is added and the mixture is vigorously shaken with 10 ml carbon tetrachloride for about 1 minute to extract the thallium(III) complex. The absorption maximum of the complex is at 426 nm in carbon tetrachloride solution. Bismuth(III) interferes with the procedure.[69]

Diethyldithiocarbamate has also proved suitable for the *determination of niobium*. By means of this reagent niobium can be precipitated quantitatively at about pH 4–5 in the presence of tantalum, titanium, zirconium, vanadium and rare earth metals. Separation of niobium from tantalum is satisfactory only if the latter is present in not more than five-fold amount.[192] The diethyldithiocarbamate complex of niobium can be extracted with chloroform at pH 5.[180]

Kreimer and Butylkin[255] have developed an interesting selective analytical application of diethyldithiocarbamate. They pointed out that only copper, mercury and silver are capable of substituting equivalent amounts of lead in the chloroform solution of the lead diethyldithiocarbamate complex. However, of these only the copper complex is coloured, and therefore the lead diethyldithiocarbamate complex may be considered a specific reagent of copper. With this method these authors successfully determined micro amounts of copper in nickel and various alloys.

Diethyldithiocarbamate has also proved suitable for the enrichment and isolation of traces of metals. For instance, the metal content of brine or the washings obtained in the purification of contaminated air can be extracted with a solution of diethyldithiocarbamate in ethyl propionate.[388] The ethyl propionate solution can be used directly for atomic absorption spectroscopic measurement. The sensitivity of the measurement is much higher in this medium than in water.

Diethylammonium diethyldithiocarbamate

$$\begin{array}{c} H_5C_2 \\ \diagdown \\ N-C \\ \diagup \diagdown \\ H_5C_2 S^- \end{array} \quad \begin{array}{c} C_2H_5 \\ \diagup \\ ^+NH_2 \\ \diagdown \\ C_2H_5 \end{array}$$

Molecular weight: 222.49
Solubility: less soluble in water, readily soluble in chloroform and carbon tetrachloride. It decomposes in aqueous solution, but is stable in organic solvents.

In water–chloroform mixtures the reagent is found quantitatively in the aqueous phase at pH > 8, while it remains quantitatively in the organic phase at pH < 4.

The reagent reacts with essentially the same metal ions as does the sodium salt of diethyldithiocarbamate.[221] However, it possesses the advantageous feature that some metal ions can be extracted with it from more acidic solutions, which in some cases may result in an increase of analytical selectivity.

The analytical applicability of several substituted dithiocarbamates has also been examined. Hulanicki[213] demonstrated the effects of various substituents in the reagent upon the properties of the corresponding complexes. In the main the majority of analytical problems can be solved equally well with diethyldithiocarbamate itself. Substitution of the reagent enhances the efficiency of the ligand in some special cases only.

TABLE 25. USE OF SODIUM AND DIETHYLAMMONIUM-DIETHYLDITHIOCARBAMATES FOR METAL ION EXTRACTION

Metal ion	pH of the aqueous phase	
	extracted with the sodium salt	extracted with the diethylammonium salt
Bi(III)	4–11	10 N H_2SO_4–pH 12
Co(III)	4–11	2.5–12
Cu(II)	4–11	10 N H_2SO_4–pH 12
Ni(II)	5–11	2.5–10
Sn(IV)	4– 6.2	unextractable
Te(IV)	4– 8.8	0–8.5
Tl(III)	4–11	5 M H_2SO_4–pH 12
V(V)	3– 6	4.0–5.5

12. FURTHER SULFUR-CONTAINING REAGENTS

Rubeanic acid

(Dithiooxamide)

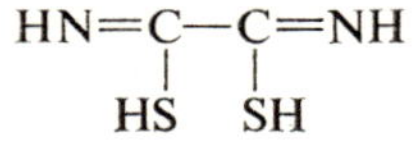

$$HN=C-C=NH$$
$$||$$
$$HSSH$$

Molecular weight: 120.19
Orange crystalline compound.
Solubility: sparingly soluble in water and ethanol, soluble in concentrated sulfuric acid to give a red solution.

Rubeanic acid may be used for the quantitative precipitation of copper, nickel and cobalt. The copper complex is dark green, the cobalt brownish-red and the nickel violet. All these complexes are chelates of outstanding stability, and this, together with their vivid colours, affords the reaction great analytical sensitivity.

The copper complex is the most stable of the three; it can be precipitated even from mild mineral acidic solution, while precipitation of the corresponding cobalt and nickel complexes requires buffered media.[371] On the other hand, palladium, platinum and silver react with rubeanic acid to give precipitates in strongly acidic solution.[481] It should be mentioned, however, that in the case of platinum and some other metals (zinc, cadmium, silver, lead, mercury) the product is the corresponding sulfide rather than the rubeanate complex.

Rubeanic acid reacts with ruthenium to give a blue water-soluble complex ion, which is utilized for the spectrophotometric determination of ruthenium.[18]

Rubeanic acid has similarly been used for the spectrophotometric determination of osmium.[469]

It has also proved suitable for the solvent extraction separation and determination of thallium.[422] It has been found that thallium exists in the organic phase in the form of a complex of composition Tl/dto/Hdto, where dto represents the rubeanate anion with one negative charge.

Thionalide

(Thioglycollic acid-β-aminonaphthalide)

Molecular weight: 217.28
White or pale yellow needles.
Solubility: sparingly soluble in water, readily soluble in organic solvents. The solubility in water is considerably increased by addition of relatively small amounts of ethanol or acetic acid.

Thionalide reacts with a variety of metal ions to give chelate complexes which are sparingly soluble in water. The difference in stability of complexes formed by the reagent with various metal ions in some cases assures a certain analytical selectivity.[50]

1. Copper(II), silver(I), mercury(II), bismuth(III), arsenic(III), tin(IV), gold(I), platinum(IV) and palladium(II) give precipitates with the reagent in acidic media (mineral acids).

2. The copper(II), mercury(II), cadmium(II), thallium(I) and gold(I) complexes precipitate from alkaline solutions containing tartrate.

3. Thallium(I), antimony(III) and bismuth(III) complexes are precipitated by the reagent from solutions containing cyanide and tartrate.

4. The thallium(I) complex also precipitates from alkaline (sodium hydroxide) solution containing cyanide and tartrate.

The above classification shows that the reaction can be made specific for thallium by appropriate choice of the pH of the solution and application of suitable masking agents. Similarly, the selectivity of the precipitation reaction can be increased for other ions too, by application of other masking agents. For instance, tin(IV) can be masked selectively in the presence of arsenic(III) and antimony(III) with phosphoric acid, while these latter two ions can be precipitated with thionalide.

The analytical application of thionalide is stimulated by the fact that its metal complexes are of stoichiometric composition, and can be dried at 105–110°C and weighed directly.

The reagent may also be utilized in volumetric analytical procedures. The thionalide content, which is equivalent to the metal to be determined, can be titrated directly iodometrically, according to the equation:

$$2 C_{10}H_7-NH-CO-CH_2-SH + I_2 =$$
$$= C_{10}H_7-NH-CO-CH_2-S-S-CH_2-CO-NH-C_{10}H_7 + 2HI.$$

Preparation of the reagent solution. A 1–2% solution of thionalide in alcohol, acetone or glacial acetic acid is used. Since thionalide readily undergoes oxidation in solution, only freshly prepared reagent solutions are used in analytical practice.

Determination of thallium(I). The sample solution containing 25–100 mg thallium is neutralized with 2 M sodium hydroxide solution; 2 g sodium tartrate and 3–5 g potassium cyanide, depending upon the concentration of the accompanying ions, are dissolved in it, it is basified by addition of 20 ml 2 M sodium hydroxide and diluted with water to 100 ml. The thallium is precipitated with 20 ml 2% thionalide solution in acetone. The mixture is heated to boiling, allowed to cool slowly, and the precipitate is collected on a G4 sintered glass filter and washed with cold water until the washings are free of cyanide ions. (Detection of cyanide in the washings is performed with iron salts (Prussian blue test).) The precipitate is then washed with a little acetone to remove excess reagent, dried at 100°C and weighed. Gravimetric factor: 0.4860.

The precipitate may be dissolved in a mixture of 3 parts glacial acetic acid and 1 part M sulfuric acid. In order to measure the thionalide concentration of the solution, which is equivalent to the thallium concentration to be determined, excess 0.02 N iodine–potassium iodine standard solution is added, and the excess iodine is titrated after a few minutes with 0.02 N sodium thiosulfate standard solution using starch indicator. 1 ml 0.02 N iodine solution corresponds to 4.09 mg thallium.

Thioglycollic acid anilide

Molecular weight: 167.21
White needles. Melting point 110.5–111°C.
Solubility: readily soluble in ethanol and diethyl ether, insoluble in benzene.

The reagent has essentially the same features as those of thionalide discussed in the foregoing. An important exception is its reaction with cobalt in ammoniacal solution, when a reddish-brown precipitate is formed which is insoluble in 2 N hydrochloric acid. The reaction is utilized for the specific detection of cobalt in the presence of metals of the sulfide group. Dilution limit: 1:10,000,000.[53] The reagent has almost no significance in quantitative analysis.

Thioacetamide

$$CH_3-\underset{\underset{S}{\parallel}}{C}-NH_2$$

Molecular weight: 76.13
White or pale yellow crystalline substance, melting point 112–114°C.
Solubility: soluble in water and ethanol, sparingly soluble in diethyl ether.

The significance of the reagent originates from its ability to take the place of hydrogen sulfide gas in analytical practice: it undergoes hydrolysis in aqueous medium to give hydrogen sulfide according to the equation:

$$CH_3CSNH_2 + 2\ H_2O = CH_3COONH_4 + H_2S.$$

In acidic media, it reacts with the cations of the first and second groups of Fresenius' qualitative analytical system, while in alkaline solutions it reacts with the members of the third group of this system to give sulfide precipitates.[29] In analytical practice a 2% aqueous solution of the reagent is usually used.

13. ARSONIC ACIDS

Phenylarsonic acid

Molecular weight: 202.02
Solubility: 3.2% in water and 15.5% in ethanol.

Arsonic acids (arsonic acid, phenylarsonic acid, p-hydroxyphenylarsonic acid, arsanilic acid) react with the tetravalent metals of group IV of the periodic system to give water-insoluble complexes of composition MA_2. The composition of the precipitates is not exactly stoichiometric, and hence the precipitates are not suitable for direct gravimetric analysis. Instead, the precipitates are transformed by ignition to the corresponding metal oxides and weighed as such. The main advantage of the application of these reagents is the possibility offered of the selective determination of zirconium(IV), hafnium(IV) and titanium(IV) in the presence of many other metals, such as zinc, manganese, nickel, cobalt, aluminium, copper, calcium, magnesium and chromium. In practice they are utilized most frequently for the determination of zirconium.

In strong mineral acidic media phenylarsonic acid forms water-insoluble precipitates with zirconium and hafnium only. In highly concentrated acetic acid solution the thorium phenylarsonate complex is also precipitated, which makes possible the separation of thorium from rare earth metals.

Preparation of the reagent solution. 2.5 g finely powdered phenylarsonic acid is dissolved in 100 ml cold distilled water and the solution is filtered.

Determination of zirconium. The sample solution containing about 50 mg zirconium in 200 ml, and which is not stronger than 4% in hydrochloric acid, is treated at room temperature with 50 ml reagent solution, the mixture is then heated to boiling, kept at this temperature for about 5 minutes and filtered through a filter paper. The precipitate is washed with 50 ml 1% hydrochloric acid solution, dried at about 120°C and ignited in an efficient hood to give zirconium dioxide. Gravimetric factor: 0.741.

The procedure gives satisfactory results in the presence of alkaline earth metals, iron, aluminium, manganese, cobalt, nickel, copper, zinc and bismuth.[376]

Arsanilic acid

(*p*-Aminophenylarsonic acid)

$$H_2N-C_6H_4-As(\rightarrow O)(O-H)_2$$

Molecular weight: 217.04
White crystalline powder. Melting point 232°C.
Solubility: sparingly soluble in water, soluble in concentrated mineral acids, alkali carbonate solutions, alcohol and diethyl ether. Practically insoluble in acetone, benzene and chloroform.

Arsanilic acid reacts essentially analogously to phenylarsonic acid. According to Chandelle,[111] in the presence of relatively large amounts of accompanying ions the determination of zirconium gives more accurate results with this reagent than with phenylarsonic acid. This statement has not yet been confirmed. Nevertheless, the reagent is much more widely used than any other arsonic acid derivative.

Preparation of the reagent solution. An aqueous solution of the sodium salt of arsanilic acid is used as reagent solution; this is always freshly prepared by reacting equivalent amounts of arsanilic acid and sodium carbonate.

Determination of zirconium. 200 ml sample solution, containing zirconium equivalent to 30–40 mg zirconium dioxide, and preferably about 0.5 N (but anyhow not more than 4 N) in sulfuric acid, is treated with reagent solution equivalent to 1 g arsanilic acid. The mixture is refluxed for 10–15 minutes, kept for 6 hours and filtered through a filter paper. The precipitate is washed three times with 5 ml 0.5 N sulfuric acid and with 100 ml cold water in portions, dried at 120°C and ignited to zirconium dioxide. Nickel, cobalt, aluminium, chromium, zinc, manganese, copper and magnesium do not interfere with the determination. Iron and titanium are disturbing. Iron can be removed by extraction with diethyl ether from 6 N hydrochloric acid solution. The removal of titanium may be achieved by oxidation with hydrogen peroxide to titanium dioxide in the presence of ammonium hydroxide.

Arsenazo I (Uranone)

Sodium 2-(2-arsonophenylazo)-1,8-dihydroxynaphthalene-3,6-disulfonate

Molecular weight: 592.30
Dark red powder.
Solubility: quite soluble in alkaline aqueous media, less soluble in acids.

The reagent reacts with many metal ions, such as uranium(IV), thorium(IV), zirconium(IV), scandium(III), lanthanum(III), cerium(III), aluminium(III), beryllium(III), titanium(III), niobium(III), tantalum(III), vanadium(IV), tin(IV), bismuth(III), gallium(III), copper(III), palladium(II), magnesium(II) and calcium(II) to give coloured complexes. It has been used for the photometric determination of the majority of the above-listed metal ions.[261, 401] Though the reagent reacts with such a variety of metal ions, the complex formation reaction may be made selective for individual metal ions by appropriate choice of the reaction conditions, primarily of the pH of the solution. For instance, thorium can be determined with Arsenazo I in the presence of almost all rare earth metals, and similarly zirconium and thorium in the presence of uranium, etc.

The stabilities of the metal complexes of Arsenazo I are lower than those of the corresponding complexes of EDTA. It can also therefore be used advantageously as indicator in complexometric determination of some metals (e.g. plutonium, thorium).[120, 331]

The analytical sensitivity of the formation of Arsenazo I complexes is comparatively high (0.05–0.1 γ/ml). However, the importance of the reagent is decreasing in analytical practice, in parallel with an increase in significance of the related reagent, Arsenazo III, which gives more stable complexes with higher absorbances, and which is therefore of greater analytical sensitivity.[402]

Arsenazo III

2,7-Bis-(2-arsonophenylazo)-1,8-dihydroxynaphthalene-3,6-disulfonic acid

Molecular weight: 776.38
Dark red, sometimes black powder.
Solubility: sparingly soluble in aqueous acidic media, readily soluble in aqueous

alkaline solutions. Insoluble in acetone, ethanol, and diethyl ether. The colour of its 0.01–0.1% aqueous solution depends upon the pH (pink at pH <4, violet at pH >5, and green in concentrated sulfuric acid).

Arsenazo III reacts with more than 30 metal ions to give coloured complexes.[402] So far it has been utilized for the determination and detection of about 25 metal ions (including 16 rare earth metals). The majority of the analytical procedures are spectrophotometric. Of these, the most important ones are summarized in Table 26.

TABLE 26. USE OF ARSENAZO III FOR SPECTROPHOTOMETRIC DETERMINATION OF METALS[81, 402]

Metal	Metal complex		Acidity of the solution examined, pH
	wave-length of absorption maximum, nm	molar extinction coefficient	
Al	585	3.75×10^3	2.9
Be	600	3.12×10^4	6.8
Ca(II)	655	10^4	6.5
Cd	540	1.20×10^3	6.3
Ce(III)	665	4.70×10^4	3.0
Hf(IV)	665	9.50×10^4	9 M HCl
Hg(II)	510	5.75×10^3	3.6
La(III)	665	4.5×10^4	3.0
Lanthanides	665	$\sim 4.50 \times 10^4$	3.0
Mg	540	2.75×10^3	6.8
Np(IV)	665	1.25×10^5	4–6 M HNO_3
Pa(IV)	660	2.20×10^4	3.5 M H_2SO_4
Pb(II)	665	10^4	5.0
Pu(IV)	670	1.36×10^5	4–8 M HNO_3
Sc	660	5.60×10^3	1.2
Se(III)	675	1.80×10^4	1.7
Sr	640	7.25×10^3	6.3
Th(IV)	665	1.30×10^5	8 M HCl
Ti(IV)	420	1.40×10^3	2.0
U(IV)	670	1.30×10^5	4 M HCl
UO_2^{2+}	665	5.30×10^4	2.0
Y(III)	665	5.50×10^4	3.0
Zn	520	2.50×10^3	6.3
Zr(IV)	665	1.20×10^5	9 M HCl

The absorption spectra of the metal complexes of Arsenazo III possess two maxima at about 610 and 665 nm. Since the absorption of the reagent is very low at 665 nm, but is considerable at 610 nm, the extinction of the complexes at 665 nm is utilized for analytical purposes. At this wavelength the molar extinction coefficients of the complexes lie in the range 5×10^4–1.3×10^5. This explains the extremely high analytical sensitivity of the reagent. (Under optimum conditions the sensitivity may reach 0.01 γ/ml.)

The main advantage of Arsenazo III over Arsenazo I lies in the considerably higher stabilities of the complexes of the former. This makes possible the analytical utilization of the reagent in strong mineral acidic media. It gives complexes with tri- and tetravalent metal ions in acidic media where hydrolysis and the consequent formation of polynuclear species do not interfere with the analytically utilized reaction. Those metals whose complexes form only in solutions of higher pH do not interfere with the analytical determination of the metal ions giving complexes in strongly acidic media (thorium, zirconium, uranium, plutonium, neptunium, etc.). This is the reason of the relatively high analytical selectivity of Arsenazo III.[400] Description for some of the more important, practically tested procedures are given below.

Determination of thorium. Arsenazo III is a specific reagent for thorium in about 3 M hydrochloric acid solutions and in the presence of oxalate as masking agent. Aluminium, niobium, tantalum, titanium, tungsten, chromium, nickel, lead and cobalt do not interfere with the determination even in thorium/accompanying metal ratios of 1:5000. Oxalic acid also eliminates interference by zirconium. The thorium complex is green. In photometric determinations the extinction of the complex is measured at 665 nm, the absorption maximum.[403]

Determination of uranium. Uranium is determined by means of Arsenazo III most advantageously in the form of its uranium(IV) complex. The uranium content of the sample, dissolved in some appropriate manner, is therefore reduced with ascorbic acid to uranium(IV). The formation of the uranium(IV) complex proceeds quantitatively in 4 M hydrochloric acid solution. The interfering effect of zirconium is eliminated with oxalic acid as masking agent. The extinction of the uranium(IV) complex is measured at 660 nm, the absorption maximum.[274]

Determination of zirconium. Only thorium(IV), uranium(IV), oxalate and fluoride interfere with the determination of zirconium with Arsenazo III. Due to its extremely high stability, the zirconium complex forms even in 9 M hydrochloric acid solution.[196]

Trace amounts of zirconium can be enriched before determination by coprecipitation with titanium hydroxide. The titanium hydroxide precipitate containing the zirconium content of the sample is dissolved in 6 M hydrochloric acid solution and the zirconium determined photometrically with Arsenazo III. Thousand-fold amounts of titanium do not interfere with the determination of zirconium. The absorption maximum of the zirconium complex of Arsenazo III is at 665 nm.[196]

Determination of scandium. The sample solution containing 5–30 γ scandium, 0.8 ml M hydrochloric acid and 0.8 ml 0.07% Arsenazo III reagent solution are combined in a 10 ml volumetric flask, the mixture is diluted to 10 ml and the absorption of the solution is measured at 680 nm in a 1 cm cell. Aluminium, iron(II), manganese(II), alkali metals and sulfate do not interfere in amounts up to 20 mg. Thorium, zirconium, uranium, bismuth, copper, calcium and iron interfere.[261, 404]

14. KALIGNOST

(Sodium tetraphenylborate)

$$\left[\underset{\text{B}}{(C_6H_5)_4} \right]^- Na^+$$

Molecular weight: 342.23
White powder.
Solubility: relatively well soluble in water.
A 0.1 M aqueous solution is generally used as reagent.

Kalignost (sodium tetraphenylborate) is a selective reagent for potassium, which can thus be determined accurately in the presence of large amounts of sodium and lithium.[368] The reagent gives precipitates with thallium(I), mercury(I), rubidium, cesium and ammonium also, and these ions must therefore be removed from the solution before determination of the potassium. Alkaline earth metals in large amounts also interfere, but these can be precipitated with sodium carbonate. The potassium can be measured directly in the alkaline filtrate.

Determination of potassium. The sample solution containing about 100 mg potassium in 100–200 ml is acidified with acetic acid to about pH 4–5 and treated with 0.1 M kalignost reagent solution so that the concentration of excess reagent in the solution is about 0.2%. After a few minutes the precipitate is collected on a sintered glass filter, washed with a 0.1% solution of kalignost in 1% acetic acid, dried at 120°C and weighed. Gravimetric factor: 0.1091.

The precipitate dissolves quantitatively in acetone. The tetraphenylborate anion, which is equivalent to the potassium to be determined, can be measured by argentometric titration with variamine blue acetate or eosin indicators. Since the solubility of the silver tetraphenylborate precipitate is significant if too much acetone is used, the potassium tetraphenylborate precipitate is dissolved in the minimum amount of acetone necessary.[149, 384]

Kalignost can also be measured by total oxidation with chromium(VI) in concentrated sulfuric acid medium.[410] The main advantage of the procedure is the low equivalent weight assured by the total oxidation. The equivalent weight of a kalignost ligand is 1/120 of the molecular weight. Accordingly, 1 ml 0.1 N sodium dichromate standard solution in concentrated sulfuric acid is equivalent to 32.56 γ potassium.

Wendlandt[475] used kalignost for the determination of thallium(I). Alimarin and Krausz[6] utilized the low solubility of the kalignost complex of thorium for the gravimetric and volumetric determination of micro amounts (300–2000 γ) of thorium. In the volumetric determination the thorium was precipitated with a standard solution of kalignost and the excess reagent was measured argentometrically.

Various derivatives of kalignost have been successfully applied for the determination of other alkali metal ions. For instance, triphenylcyanoborate anions give precipitates only with cesium of the alkali metals, while tetra-p-tolylborate reacts with all alkali ions but lithium to give precipitates. Thus by appropriate combination of the various kalignost derivatives the separation of alkali metal ions can be achieved.[39]

CHAPTER 3

SUMMARIZING TABLES

SOME MORE IMPORTANT ORGANIC REAGENTS AND THEIR SCOPE OF APPLICATION

During the past 50–60 years several hundred organic reagents used in metal analysis have been described in the analytical chemistry literature. The scope of application of these reagents is extremely variable. Many of them, particularly those described earlier, have been completely replaced in analytical practice by newer, more efficient reagents. There are also many which are useful only for the simple accomplishment of some special, unique problems. Thus, although being of great value, these reagents find application only within narrow limits.

In the author's experience, the great majority of problems encountered in metal analytical practice that are soluble by means of organic reagents can generally be dealt with by one or other of the 38 reagents discussed in the previous part of this book. None the less, it appears necessary to present about a further one hundred organic reagents together with an indication of their fields of application. In some cases these may replace those discussed in more detail in the foregoing. In special cases they may also be more effective than those standard reagents. A more detailed discussion of these reagents would be of value if the size of this book permitted it. Tables 27 and 28 summarize the reagents in alphabetical order of the reagents and the metals, respectively.

Table 27 shows what metal ions can be determined with a given reagent and by what method.

Table 28 provides information on the reagents and methods suitable for the determination of individual metals.

TABLE 27. SOME IMPORTANT ORGANIC REAGENTS AND THEIR FIELDS OF APPLICATION

Reagent	Chemical name	Molecular weight	Measurable metal ion	Method of measurement
Alberon	5,5′-dimethyl-4′-hydroxy-2″,6″-dichloro-3″-sulfo-fuchson-3,3′-dicarboxylic acid	539.35	Be^{2+}, Al^{3+}	photometric
Alizarin	1,2-dihydroxyanthraquinone	240.22	Al^{3+}, Zr^{IV}, Th^{IV}	photometric
Alizarin S Sodium alizarin sulfonate	1,2-dihydroxyanthraquinone-3-sulfonate sodium salt	360.28	Al^{3+}, Zr^{IV}, Th^{IV}, Ga^{3+}, Ti^{IV}	photometric

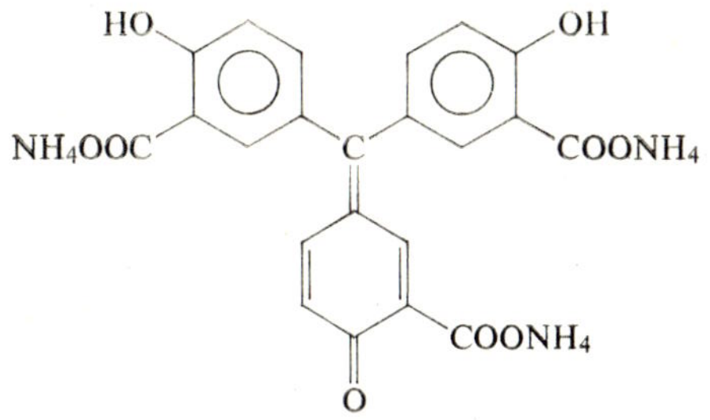 aurine-3,3′,3″-tricarboxylic acid triammonium salt	473.45	Al^{3+}	photometric
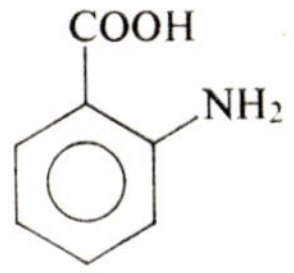 o-aminobenzoic acid	137.14	Cd^{2+}, Co^{2+}, Cu^{2+}, Hg^{2+}, Mn^{2+}, Ni^{2+}, Pb^{2+}, Zn^{2+}	gravimetric
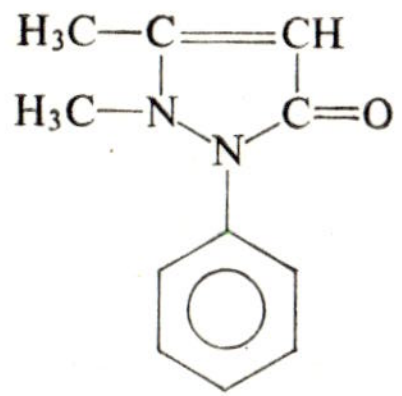 1-phenyl-2,3-dimethyl-3-pyrazolin-5-one		Co^{2+}	photometric
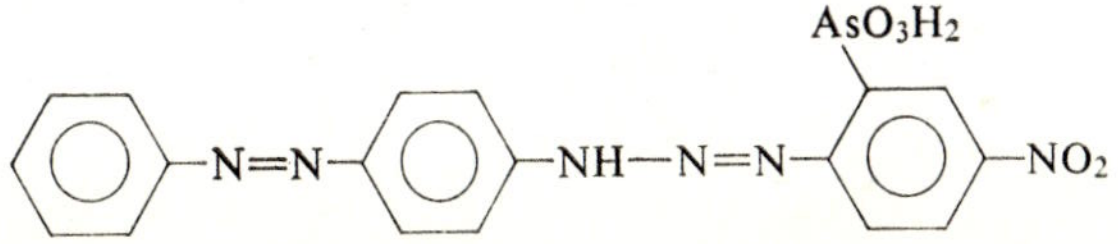 4′-phenylazo-4-nitrodiazo-aminobenzene-2-arsonic acid	470.28	Pb^{2+}, Zn^{2+}	photometric

Aluminon

Anthranilic acid

Antipyrine

Arsacen

TABLE 27 (continued)

Reagent	Chemical name	Molecular weight	Measurable metal ion	Method of measurement
Arsenazo I	2-(2'-arsonophenylazo)-1,8-dihydroxynaphthalene-3,6-disulfonic acid disodium salt	592.30	Al^{3+}, Zr^{IV}, In^{3+}, lanthanides, Y and Sc	photometric
Arsenazo III	2,7-bis(2'-arsonophenylazo)-1,8-dihydroxynaphthalene-3,6-disulfonic acid	776.38	Th^{IV}, Zr^{IV}, U^{IV}, U^{VI}, Hf^{IV}, Sc^{3+}, lanthanides, Y and Sc	photometric

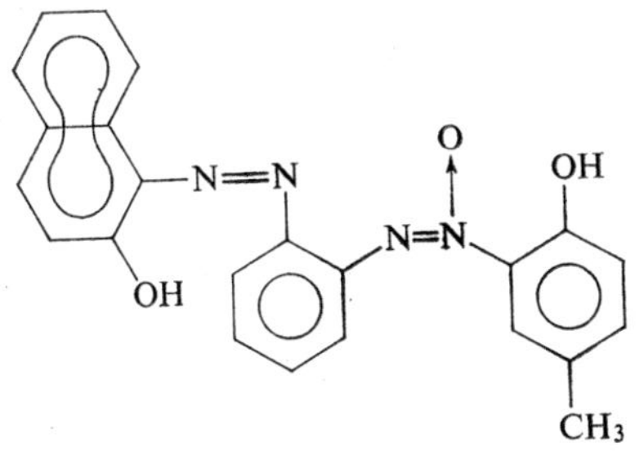 2-hydroxy-5-methyl-2'(2-hydroxy-1-naphthylazo)-azoxybenzene	398.43	Ca^{2+}	photometric
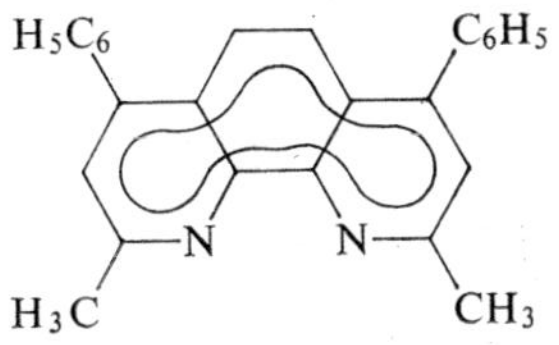 2,9-dimethyl-4,7-diphenyl-1,10-phenanthroline		Cu^{2+}	photometric
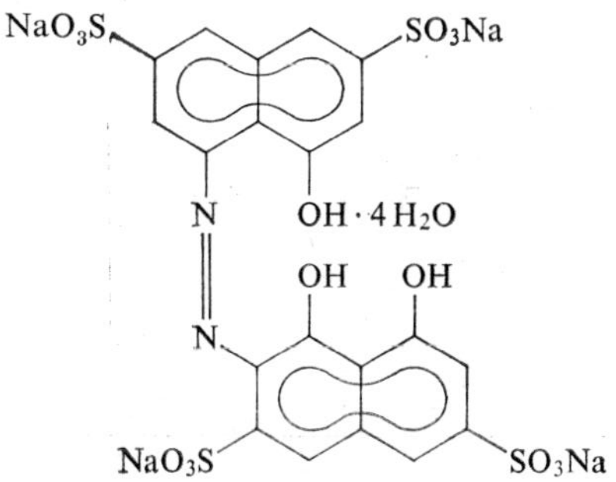 1,8,8'-trihydroxy-2,1'-azo-naphthalene-3,6,3',6'-tetra-sulfonic acid tetrasodium salt, tetrahydrate	810.58	Be^{2+}	photometric

Azoazoxy

Bathocuproine

Berillon II
IREA

TABLE 27 (continued)

Reagent	Chemical name	Molecular weight	Measurable metal ion	Method of measurement
2,2′-Bis(cinchoninic acid) potassium salt	2,2′-biquinoline-4,4′-dicarboxylic acid dipotassium salt	420.52	Cu^{2+}	photometric
Bissalicylideneethylenediamine	N,N′-bissalicylideneethylenediamine	268.32	Mg^{2+}	photometric
Brilliant Green	N,N′-diethyl-4-diethylaminofuchsonimmonium-hydrogen sulfate	482.65	Sb^V, Tl^{3+}, Zn^{2+}	photometric

SUMMARIZING TABLES

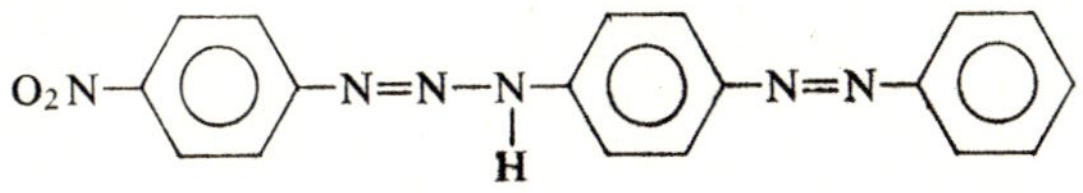 4-phenylazo-4′-nitrodiazo-aminobenzene	346.35	Cd^{2+}	photometric
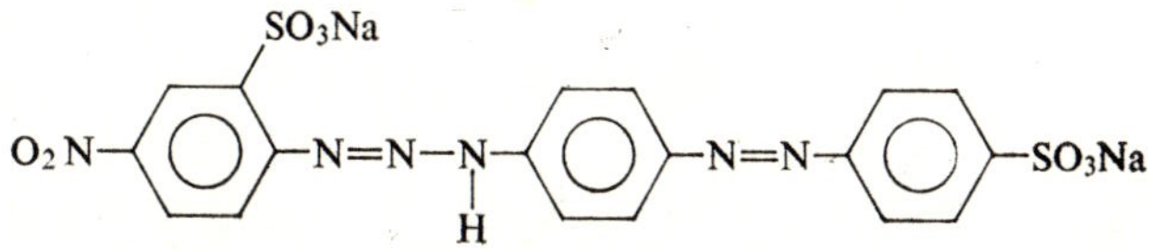 4-(p-sulfophenylazo)-2′-sulfo-4′-nitrodiazoaminobenzene disodium salt	550.46	Cd^{2+}	photometric
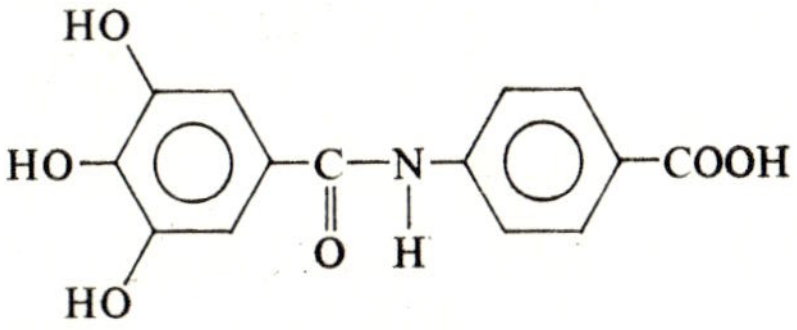 4-(3′,4′,5′-trihydroxybenzoyl-amino)-benzoic acid	289.24	Ti^{IV}	photometric
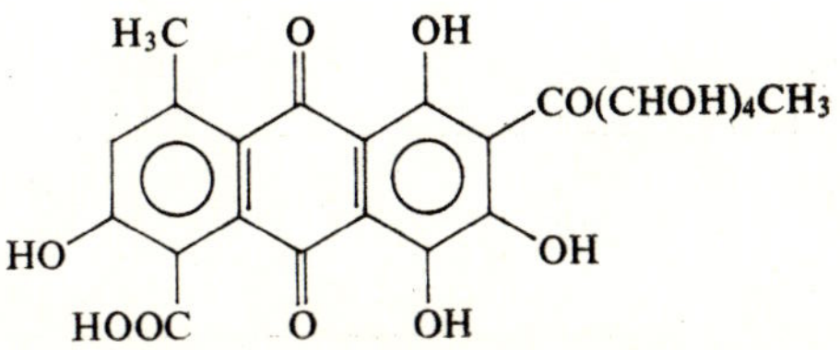 2,5,7,8-tetrahydroxy-4-methyl-6-(2′,3′,4′,5′-tetrahydroxy-hexanoyl)-anthraquinone-1-carboxylic acid		B	photometric

Cadion

Cadion IREA

p-Carboxygallanilide

Carminic acid

TABLE 27 (continued)

Reagent	Chemical name	Molecular weight	Measurable metal ion	Method of measurement
Chloranilic acid	2,5-dihydroxy-3,6-dichloro-p-benzoquinone		Al^{3+}, Ca^{2+}, Mo^{VI}, Pb^{2+}, Sr^{2+}	photometric
Chrome Azurol S	5,5′-dimethyl-4′-hydroxy-2″,6″-dichloro-3″-sulfofuchson-3,3-dicarboxylic acid trisodium salt		Al^{3+}, Be^{3+}	photometric
Chromotropic acid disodium salt	1,8-dihydroxynaphthalene-3,6-disulfonic acid disodium salt	364.27	Ti^{IV}, Cr^{VI}	photometric

Cinchonine $C_{19}H_{22}N_2O$	294.19	Bi^{3+}, Ir, Pt^{IV}, Mo^{VI}, W^{VI}	gravimetric
Crystal Violet N,N-dimethyl-4,4''-bis(di-methylamino)-fuchson-immonium-chloride, mono-hydrate	570.14	Sb^V, Zn^{2+}, Tl^{3+}, Cd^{2+}, Hg^{2+}	photometric
Cupferron N-nitroso-N-phenylhydroxyl-amine ammonium salt	155.16	Bi^{3+}, Cu^{2+}, Fe^{3+}, Ga^{3+}, Nb^V, Ta^V, Ti^{IV}, Th^{IV}, V^V, Zr^{IV}, U^V	gravimetric
Cuproin 2,2'-diquinolyl	256.30	Cu^{2+}, Ti^{3+}	photometric

TABLE 27 (continued)

Reagent	Chemical name	Molecular weight	Measurable metal ion	Method of measurement
Cupron	α-benzoinoxime	227.27	Cu^{2+}	gravimetric
Curcumin	1,7-bis(3-methoxy-4-hydroxy-phenyl)-hepta-1,6-diene-3,5-dione	368.39	B^{III}, Be^{2+}	photometric
Cystein	α-amino-β-mercaptopropionic acid	121.15	Co^{2+}, Cu^{2+}, K^+	photometric

SUMMARIZING TABLES

Name	Structure	M.W.	Ions	Method
Daxime	1,3-dimethyl-5,6,-dihydro-5-oximino-6-imminouracil	184.16	Cu^{2+}, Pd^{2+}	photometric, gravimetric
Diallyldithiocarbamoylhydrazine	Hydrazine-1,2-dithiocarboxylic acid diallylamide	230.36	Cu^{2+}, Pb^{2+}, Zn^{2+}, Ni^{2+}, Ag^+	gravimetric
Diaminoanthraquinonesulfonic acid	3,4-diaminoanthraquinone-1-sulfonic acid	318.28	Cu^{2+}	photometric
Diaminobenzidine	3,4,3',4'-tetraaminodiphenyl	214.28	Se^{IV}	photometric

TABLE 27 (continued)

Reagent	Chemical name	Molecular weight	Measurable metal ion	Method of measurement
Diantipyrylmethane	bis(1-phenyl-2,3-dimethyl-5-oxo-3-pyrazolin-4-yl)-methane	388.48	Cd^{2+}, Ti^{IV}, Fe^{3+}, Bi^{3+}, Co^{2+}	gravimetric
Diazobenzenesulfonic acid		184.17	Al^{3+}, Mg^{2+}	photometric
Dibromooxine	5,7-dibromo-8-quinoline	302.97	Al^{3+}, Co^{2+}, Cu^{2+}, Ga, Ti^{4+}, Pb^{2+}, Fe^{2+}, Tl^{4+}	gravimetric

Name	Structure	Chemical name	M.W.	Ions determined	Method
Dibromooxyquinaldine		2-methyl-5,7-dibromoquinolinol	317.01	Sn^{IV}	extr.
2,7-Dichlorochromotropic acid		1,8-dihydroxy-2,7-dichloronaphthalene-3,6-disulfonic acid disodium salt, dihydrate	469.18	Ti^{IV}	photometric
Dichlorooxine		5,7-dichloro-8-quinolinol	214.05	$Ti^{4+}, Cu^{2+}, Fe^{3+}$	gravimetric
p-Dimethylaminobenzylidene-rhodanine		2-thioxo-5-(p-dimethylaminobenzylidene)-4-thiazolidinone	264.38	$Ag^+, Au^+, Hg^{2+}, Pd^{2+}, Cu^+, Pt^{IV}, CN^-$	photometric, volumetric

TABLE 27 (continued)

Reagent	Chemical name	Molecular weight	Measurable metal ion	Method of measurement
p-Dimethylaminophenylfluorone	2,6,7-trihydroxy-9-(p-dimethylaminophenyl)-fluorone	363.37	Ta^V	photometric
Dimethylglyoxime	diacetyldioxime	116.12	Ni^{2+}, Co^{2+}, Pd^{2+}, Cu^{2+}, Fe^{2+}	gravimetric, volumetric, photometric
Diphenylcarbazide	1,5-diphenylcarbohydrazide	242.29	Cr^{IV}, Hg^{2+}, Pb^{2+}, Cd^{2+}	photometric

SUMMARIZING TABLES 153

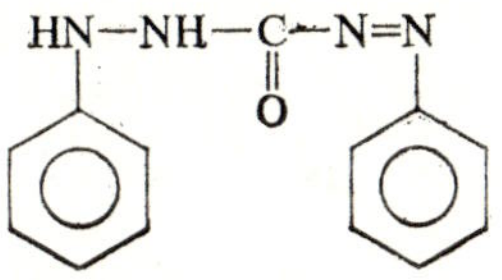 1,5-diphenylcarbazone	240.37	Hg^{2+}, Ag^+	photometric
α-benzyldioxime	240.27	Ni^{2+}	gravimetric
2,4,6,2',4',6'-hexanitrodiphenylamine	439.23	K^+, Rb^+, Cs^+	gravimetric, photometric
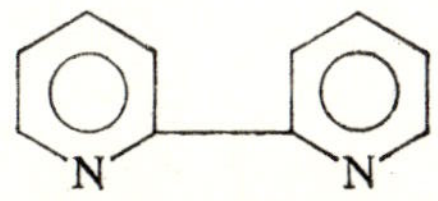 2,2'-dipyridyl	156.19	Fe^{2+}	photometric

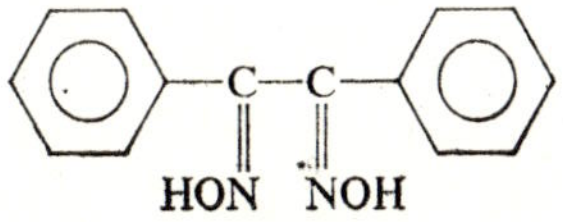

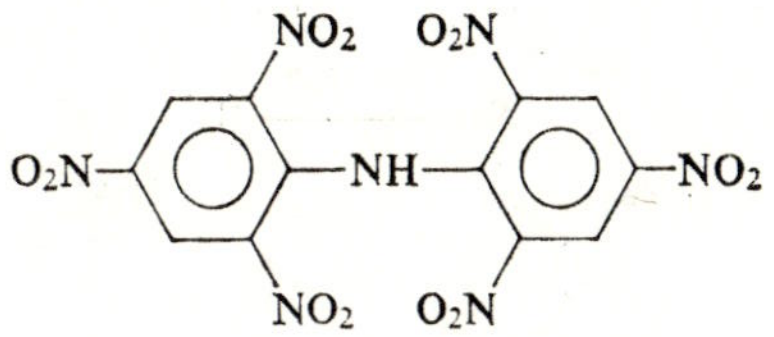

Diphenylcarbazone

α-Diphenylglyoxime

Dipicrylamine

α,α'-Dipyridyl

TABLE 27 (continued)

Reagent	Chemical name	Molecular weight	Measurable metal ion	Method of measurement
8,8′-Diquinolyldisulfide	8,8′-diquinolyldisulfide	320.44	Cu^{2+}	photometric
Dithiol	1-methyl-3,4,-dimercapto-benzene	156.27	Sn^{2+}, W^{VI}, Mo^{VI}	photometric
Dithizone	1,5-diphenylthiocarbazone	256.34	Ag^+, Bi^{3+}, Cd^{2+}, Co^{2+}, Cu^{2+}, Hg^{2+}, Pd^{2+}, Zn^{2+}, Ni^{2+}, In^{3+}, Tl^{3+}, Sn^{2+}	photometric

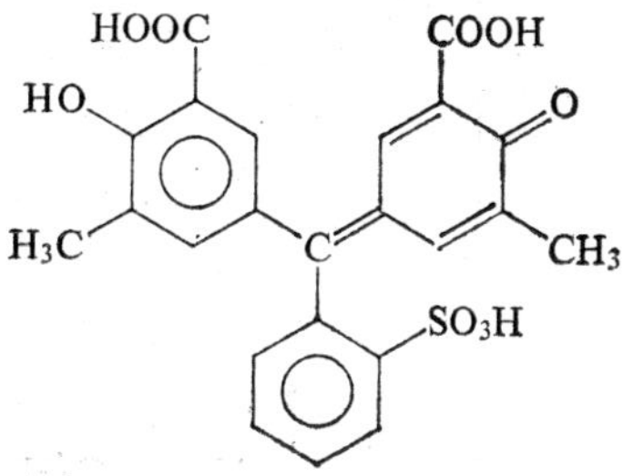 5,5′-dimethyl-4′-hydroxy-2″-sulfofuchson-3,3′-dicarboxylic acid		Al^{3+}	photometric
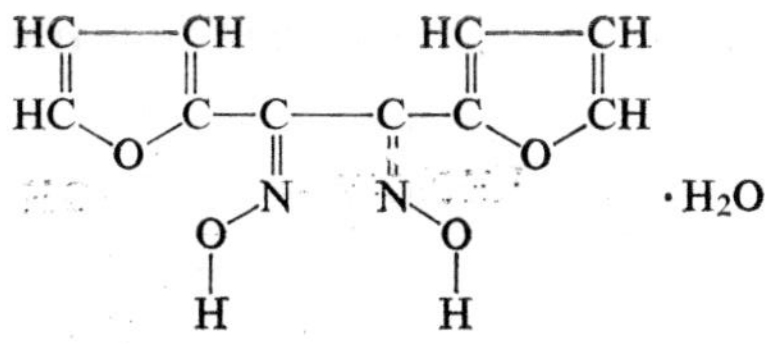	238.21	Ni^{2+}, Pd^{2+}, Pt^{IV}	gravimetric, photometric
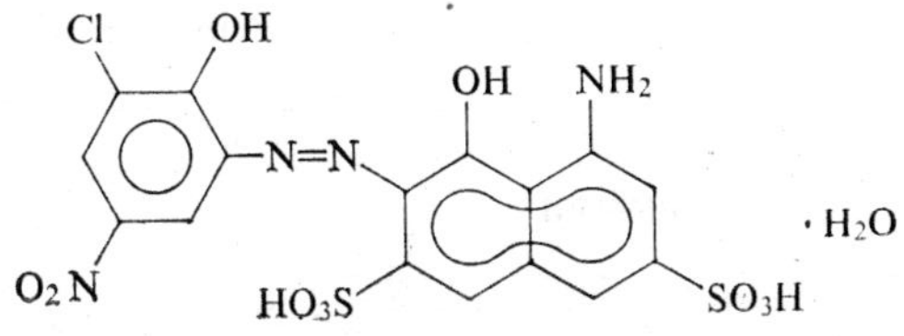 1-hydroxy-2-(2′-hydroxy-3′-chloro-5′-nitrophenylazo)-8-aminonaphthalene-3,6-disulfonic acid monohydrate	539.90	Ga^{3+}	photometric

Eriochrome cyanin

α-Furildioxime

Gallion IREA

TABLE 27 (continued)

Reagent	Chemical name	Molecular weight	Measurable metal ion	Method of measurement
Hematoxylin $C_{16}H_{14}O_6 \cdot 3\,H_2O$		356.32	Al^{3+}, Cu^{2+}, Fe^{2+}, Fe^{3+}	photometric
4-Hydroxy-3-nitrophenylarsonic acid	4-hydroxy-3-nitrophenylarsonic acid		Cd^{2+}	photometric
8-Hydroxyquinaldine	2-methyl-8-quinolinol	159.17	Zn^{2+}, Mg^{2+}	gravimetric, volumetric
Kalignost	sodium tetraphenylborate	342.24	K^+, Rb^+, Cs^+, Tl^+, NH_4^+	gravimetric, volumetric, photometric

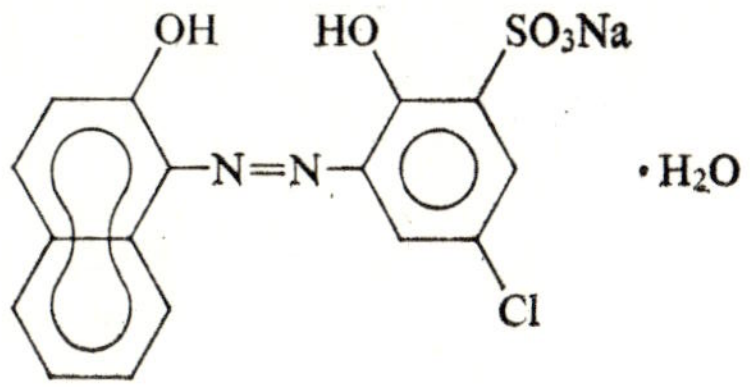 1-(2-hydroxy-3-sulfo-5-chlorophenyl-azo)-2-naphthol sodium salt monohydrate	418.80	Mg^{2+}	photometric
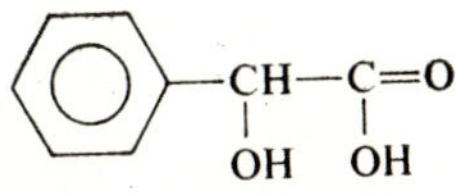 α-hydroxyphenylacetic acid	152.14	Zr^{4+}	extr. + photometric
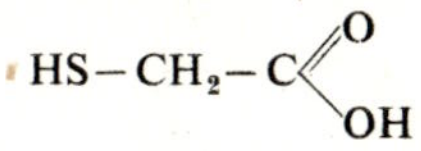	92.11	Fe^{2+}, Al^{3+}, W^{VI}, Sn^{2+}	photometric
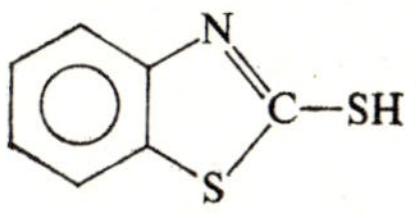	167.26	Ag^+, Au^{3+}, Bi^{3+}, Cd^{2+}, Cu^{2+}, Pb^{2+}, Tl^{3+}	gravimetric
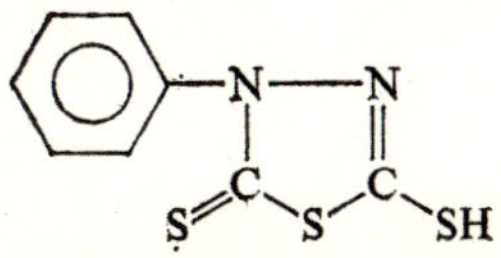 3-phenyl-5-mercapto-1,3,4-triadozol-2(3H)-thione	226.35	Bi^{3+}	photometric

Magneson IREA

Mandelic acid

Mercaptoacetic acid
$HS-CH_2-C\begin{matrix}=O\\OH\end{matrix}$

Mercaptobenzothiazole

Mercaptophenylthio-thiodiazolone

TABLE 27 (continued)

Reagent	Chemical name	Molecular weight	Measurable metal ion	Method of measurement
Morin	3,5,7,2′,4′-pentahydroxyflavone	298.26	Al^{3+}, Ga^{3+}, Zr^{IV}, Th^{IV}	photometric
β-Naphthoquinoline	benzoquinoline	179.22	Cd^{2+}	gravimetric
Neocuproin	2,9-dimethyl-1,10-phenanthroline	226.28	Cu^{2+}	extr. + photometric

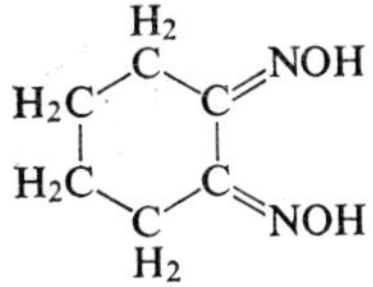 1,2-cyclohexandiondioxime	142.16	Ni^{2+}, Fe^{2+}, Pd^{2+}	photometric
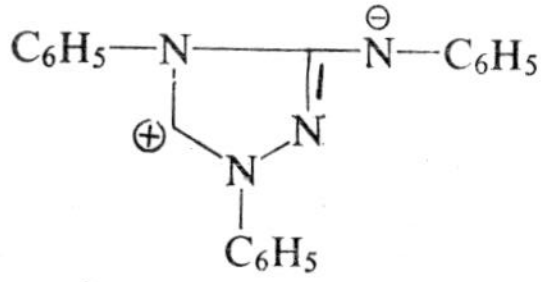 anhydro-1,4-diphenyl-3-anilino-1,2,4-triazol	312.38	ReO_4^-, NO_3^-, ClO_4^-	gravimetric
α-nitroso-β-naphthol 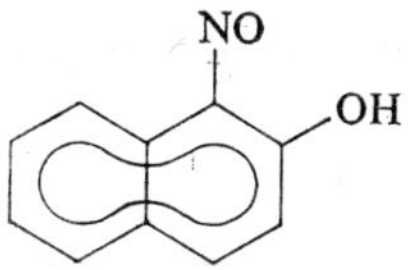	173.17	Co^{2+}, Fe^{3+}, Pd^{2+}, Cu^{2+}	gravimetric, photometric
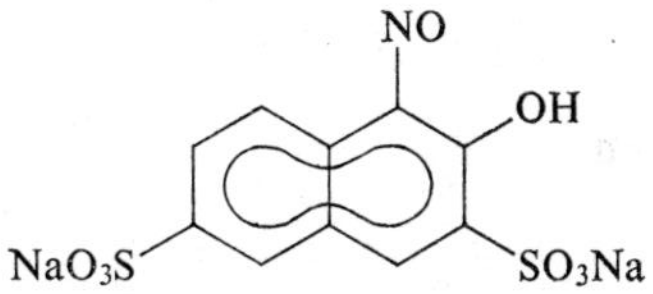 1-nitroso-2-naphthol-3,6-disulfonic acid disodium salt	377.27	Co^{2+}, K^+	photometric

Nioxime

Nitron

α-Nitroso-β-naphthol

Nitroso-R salt

TABLE 27 (continued)

Reagent	Chemical name	Molecular weight	Measurable metal ion	Method of measurement
Oxalic acid HO-C(=O)-C(=O)-OH · 2 H$_2$O		126.07	Ca^{2+}, La^{3+}	gravimetric
Oxine, 8-Hydroxyquinoline	8-quinolinol	145.16	Al^{3+}, Mg^{2+}, Zn^{2+}, Bi^{3+}, Cd^{2+}, Cu^{2+}, Co^{2+}, Ga^{3+}, In^{3+}, TiO^{2+}, Zr^{4+}, MoO_2^{2+}, WO_2^{2+}, Mn^{2+}, Fe^{3+}, Ni^{2+}, $V_2O_3^+$, Pb^{2+}, Pd^{2+}, Sb^{3+}, Cr^{3+}	gravimetric, volumetric, photometric
PAN	1-(2-pyridylazo)-2-naphthol	249.28	Co^{3+}, U^{VI}, V^V, Bi^{3+}, Nb^{IV}, Cs^{IV}, Sb^{3+}, Ta	photometric

SUMMARIZING TABLES

photometric	photometric	photometric	gravimetric
Sb^V, Zn^{2+}, Cd^{2+}, Tl^{3+}, Hg^{2+}	Fe^{2+}	Mg^{2+}	Nb^V, Ta^V, Zr^{IV}
	198.23	484.30	202.03
N,N-dimethyl-4-dimethyl-amino-4''-methylamino-fuchsonimmonium-chloride	1,10-phenanthroline monohydrate	4,4'-bis(p-hydroxyphenyl-azo)-3,3'-dinitrodiphenyl	

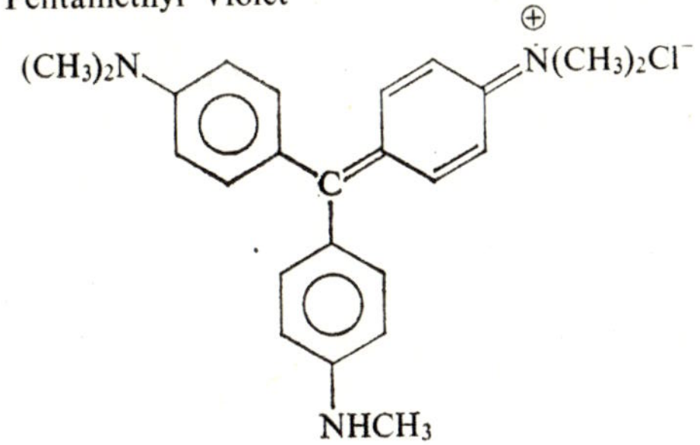

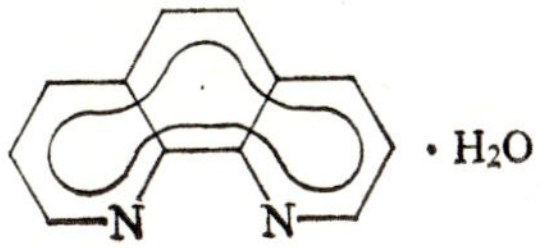

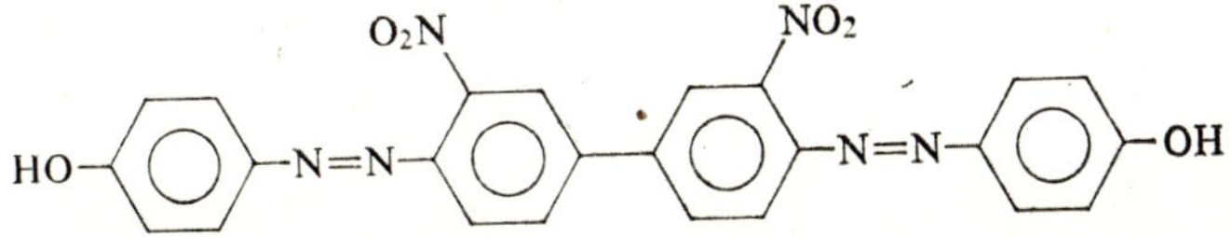

Pentamethyl Violet

1,10-Phenanthroline

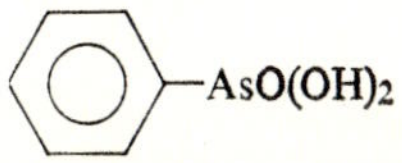

Phenazo

Phenylarsonic acid

TABLE 27 (continued)

Reagent	Chemical name	Molecular weight	Measurable metal ion	Method of measurement
Phenylfluorone	2,3,7-trihydroxy-9-phenyl-fluorone	320.31	Ga^{IV}, Sn^{IV}, Ta^{V}	photometric
Phenylthiohydantoic acid	N-phenyl-S-carboxymethyl-isothiourea	210.25	Sb^{3+}, Bi^{3+}, Cd^{2+}, Co^{2+}, Cu^{2+}, Pb^{2+}, Ag^{+}	gravimetric
Phthalic acid	benzene-o-dicarboxylic acid	166.13	Pb^{2+}	gravimetric

SUMMARIZING TABLES

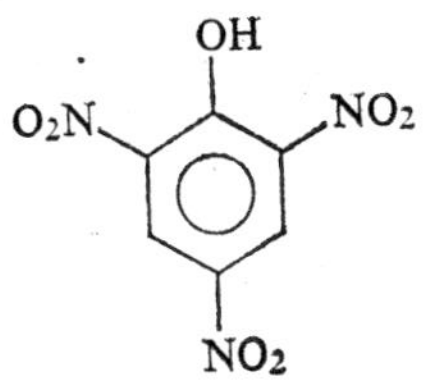 2,4,6-trinitrophenol	229.11	Bi^{3+}	gravimetric
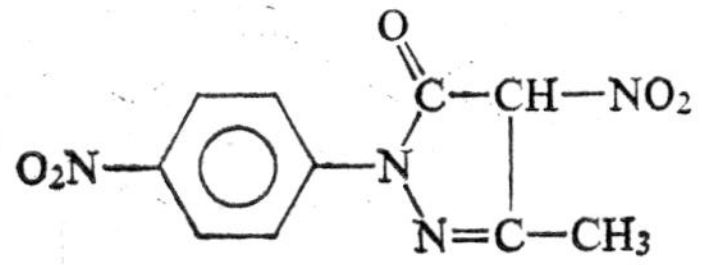 1-(p-nitrophenyl)-3-methyl-4-nitro-2-pyrazolin-5-one	264.21	Pb^{2+}, Ca^{2+}, Sr^{2+}, Mg^{2+}, Th^{IV}	gravimetric, photometric
potassium-o-ethyldithio-carbamate	160.23	Co^{2+}, Cu^{2+}, Ni^{2+}, Mo^{VI}	photometric
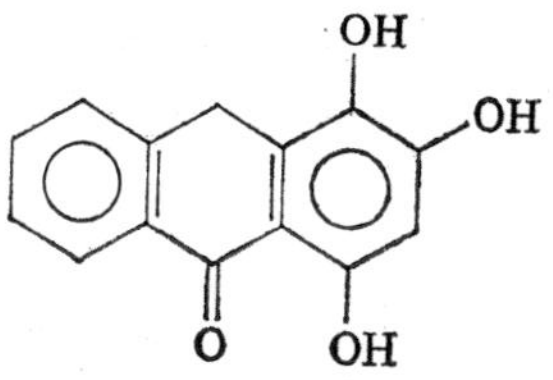 1,2,4-trihydroxy-anthra-quinone	256.22	Zr^{IV}	photometric

Picric acid

Picrolonic acid

Potassium xanthate
$C_2H_5-O-C(=S)-SK$

Purpurine

TABLE 27 (*continued*)

Reagent	Chemical name	Molecular weight	Measurable metal ion	Method of measurement
Pyridine + SCN⁻			Co^{2+}, Ni^{2+}, Zn^{2+}, Cd^{2+}	gravimetric, volumetric
Pyrocatechol Violet	3,3-bis(3,4-dihydroxyphenyl)-$3H$-2,1-benzoxathiol-1,1'-dioxide		Bi^{3+}, Cu^{2+}, Th^{4+}, Zr^{4+}	photometric
Pyrogallol	1,2,3-trihydroxybenzene	126.11	Bi^{3+}, Sb^{3+}, B^{III}, Ta^V, Nb^V	gravimetric, photometric

SUMMARIZING TABLES

Quercetin $C_{15}H_{10}O_7 \cdot 2H_2O$ 3,5,7,3',4'-pentahydroxyflavone	338.26	Al^{3+}	nephelometric
Quinaldic acid 2-quinolinecarboxylic acid, dihydrate	210.21	Cu^{2+}, Pb^{2+}, Ag^+, Mn^{2+}, Ni^{2+}, Co^{2+}, Fe^{2+}, Cd^{2+}, Zn^{2+}, UO_2^{2+}, $FeOH^{2+}$, $AlOH^{2+}$, $CrOH^{2+}$,	gravimetric
Quinalizarin 1,2,5,8-tetrahydroxyanthraquinone	272.22	Be^{2+}, B^{III}, Mg^{2+}, Al^{3+}, Ga^{3+}	photometric
Resorcylaldoxime 2,4-dihydroxybenzaldoxime	153.13	Fe^{3+}	photometric

TABLE 27 (continued)

Reagent	Chemical name	Molecular weight	Measurable metal ion	Method of measurement
Rhodamine B	N,N-diethyl-6-diethylamino-9-(o-carboxyphenyl)-fluor-immonium-chloride	479.03	Sb^V, Zn^{2+}, Ga^{3+}, Tl^{3+}, W^{VI}	photometric
Rhodamine 6G	N-ethyl-6-ethylamino-9-(o-ethoxycarbonylphenyl)-fluorimmonium-chloride	450.97	Re^{VII}, In^{3+}, Tl^+	photometric
Rubeanic acid	dithiooxamide	120.19	Co^{3+}, Cu^{2+}, Ni^{2+}, Pt^{4+}, Ru^{IV}	gravimetric

SUMMARIZING TABLES

Name	Structure	M	Ions	Method
Salicylaldoxime	2-hydroxy-C₆H₄-CH=N-OH	137.14	Cu^{2+}, Pd^{2+}, Fe^{3+}	gravimetric, photometric
Salicylic acid	2-hydroxy-C₆H₄-COOH	138.13	Fe^{3+}, Cu^{2+}	photometric
Salicylideneaminophenol	2-HO-C₆H₄-CH=N-C₆H₄-OH	213.24	Al^{3+}	photometric
Sodium diethyldithiocarbamate	$(C_2H_5)_2N-C(=S)-SNa \cdot 3 H_2O$	225.34	Cu^{2+}, Ni^{2+}, UO_2^{2+}	photometric
2-hydroxybenzoic acid				
2,2′-dihydroxybenzylidene-aniline				
sodium diethyldithiocarbamate trihydrate				

TABLE 27 (continued)

Reagent	Chemical name	Molecular weight	Measurable metal ion	Method of measurement
Sodium sulfosalicylate	2-hydroxy-5-sulfobenzoic acid sodium salt, dihydrate	254.22	Fe^{3+}, Fe^{2+}	photometric
Stilbazo	4,4′-bis-(3,4-dihydroxyphenyl-azo)-stilbene-2,2′-disulfonic acid diammonium salt	646.67	Al^{3+}	photometric

Stilbexon	4,4'-bis(N,N-di(carboxymethyl)-amino)-stilbene-2,2'-disulfonic acid disodium salt	646.51	Fe^{3+}	photometric
Sulfanilic acid	p-aminobenzenesulfonic acid	173.18	Al^{3+}, Mg^{2+}	photometric
Sulfarsacene	4'-(p-sulfophenylazo)-4-nitrodiazoaminobenzene-2-arsonic acid sodium salt	572.32	Pb^{2+}, Zn^{2+}	photometric

TABLE 27 (continued)

Reagent	Chemical name	Molecular weight	Measurable metal ion	Method of measurement
Sulfonazo	bis-(3-1'-hydroxy-8'-amino-3'6'-disulfo-2-naphthyl-azo-4-hydroxyphenyl)-sulfon	976.93	Sc^{3+}	photometric
Sulfosalicylic acid	2-hydroxy-5-sulfobenzoic acid	254.21	Fe^{3+}, Ti^{IV}	photometric

SUMMARIZING TABLES

Tannin $C_{12}H_{52}O_{46}$	170.00	Be^{2+}, Al^{3+}, Ga^{3+}, Nb^{V}, Ta^{V}, Zr^{IV}, Mo^{VI}		gravimetric
Tetraphenylarsonium chloride [(C₆H₅)₄As]⁺ Cl⁻	454.56	Cd^{2+}, Mn^{2+}, Hg^{2+}, Sn^{2+}, Zn^{2+}		volumetric
Thioacetamide $CH_3-C(=S)-NH_2$	75.13	Bi^{3+}, Al^{3+}, Sb^{3+}, Cd^{2+}, Pb^{2+}, Sn^{2+}, Hg^+, MoO_2^{2+}, Cu^{2+}, Pd^{2+}		gravimetric, photometric
Thionalid (naphthyl-NH—CO—CH₂—SH)	217.29	Tl^+, As^{III}, Sb^{III}, Sn^{2+}, Ag^+, Au^{III}, Cu^{2+}, Hg^{2+}, Pd^{2+}, Bi^{3+}, Rh^{III}, Ru^{III}, Mn^{2+}, Pb^{2+}, Ni^{2+}, Co^{2+}	mercaptoacetic acid-β-naphthylamide	photometric

TABLE 27 (*continued*)

Reagent	Chemical name	Molecular weight	Measurable metal ion	Method of measurement
Thiooxine (structure: 8-hydroxyquinoline analog with SNa group, ·3H$_2$O)	8-quinolinethiol sodium salt, trihydrate	237.25	Pd^{2+}, Cu^{2+}, Mo^{VI}, Re^{VII}, In^{3+}, Mn^{2+}, V^{IV}, Co^{2+}, Ru^{3+}, Os^{3+}, Tl$^+$	photometric
Thiophenol (structure: C$_6$H$_5$SH)			Au^{3+}	gravimetric
Thiourea H$_2$N–C–NH$_2$ $\parallel$ S		76.12	Bi^{3+}, Os^{VI}, Ru^{VI}	photometric

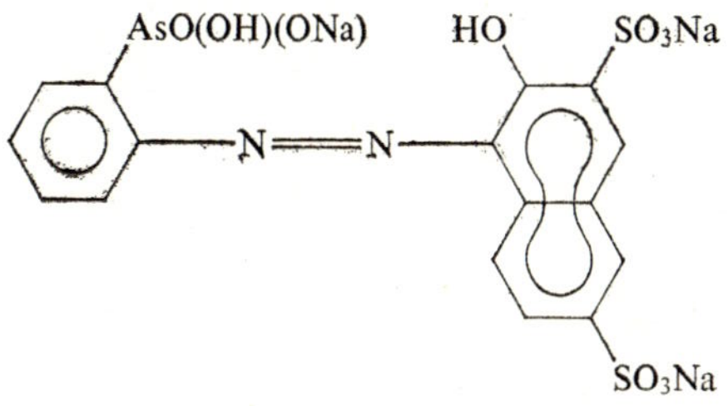 1-(o-arsonophenylazo)-2-hydroxynaphthalene-3,6-disulfonic acid trisodium salt	598.29	Th^{IV}, F^-, U^{IV}, Zn^{2+}, Zr^{IV}, Ta^V, Nb^V, lanthanides, Y and Sc, Bi^{3+}, Li^+	photometric
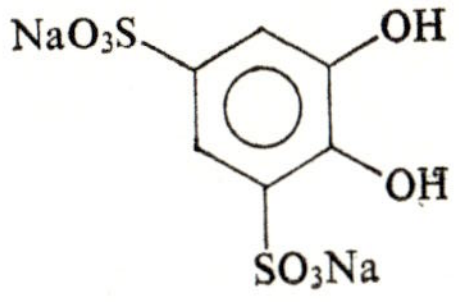 4,5-dihydroxy-1,3-benzene-disulfonic acid disodium salt	695.75	Fe^{3+}, Mg^{2+}, Ti^{IV}	photometric
p,p'-bis[6-methyl-2-benzthiazolyl]-diazoaminobenzene-o,o'-disulfonic acid disodium salt		Mg^{2+}	photometric

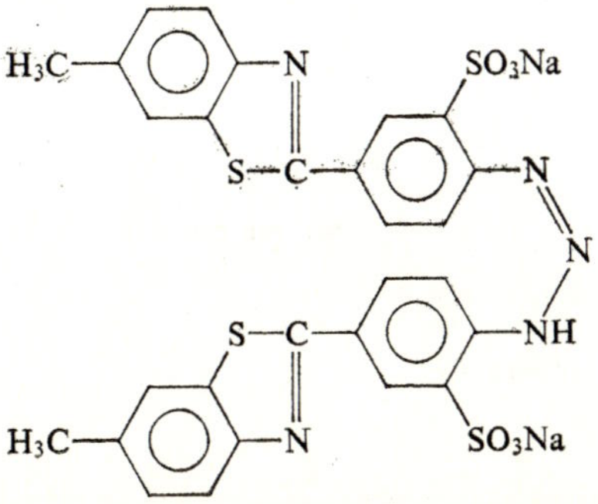

Thoron

Tiron

Titan Yellow

TABLE 27 (continued)

Reagent	Chemical name	Molecular weight	Measurable metal ion	Method of measurement
8-Tosylaminoquinoline	8-(p-toluenesulfonamido)-quinoline	298.35	Zn^{2+}, Cd^{2+}	photometric
TTA, Thenoyltrifluoroacetone	1-tienyl-4,4,4,-trifluoro-1,3-butanedione	222.2	Ce^{3+}, Co^{2+}, Cr^{3+}, Cu^{2+}, Fe^{3+}, U^{VI}, V^V	extr. + photometr.

Xylenol Orange	3,3-bis[2-methyl-3-hydroxy-4-di(carboxymethyl)-aminophenyl]-3H-2,1-benzoxathiol-1,1-dioxide		Bi^{3+}, Ga^{2+}, Zn^{2+}, Zr^{4+}	photometric
Zirconon	2′-hydroxy-5′-methylazo-benzene-4-sulfonic acid	292.32	Zr^{IV}	gravimetric

TABLE 28. ORGANIC REAGENTS USED FOR DETERMINATION OF METAL IONS

Element	Reagent	Method
Aluminium	Alberon	photometric
	Alizarin	photometric
	alizarinsulfonic acid sodium salt	photometric
	Aluminon	photometric
	Arsenazo	photometric
	chloranilic acid	photometric
	Chrome Azurol S	photometric
	Cupferron	photometric
	diazobenzenesulfonic acid	photometric
	Eriochrome Cyanin	photometric
	hematoxylin	photometric
	8-hydroxyquinoline	gravimetric, volumetric
	mercaptoacetic acid	photometric
	morin	photometric
	pyridine	photometric
	quercetin	photometric
	quinaldic acid	photometric
	quinalizarin	photometric
	salicylideneaminophenol	photometric
	Stilbazo	photometric
	sulfosalicylic acid	photometric
	tannin	gravimetric
Ammonia	Kalignost	gravimetr., volumetr., photometr.
Antimony	pyrogallol	gravimetric
	pyrrolidinedithiocarboxylic acid disodium salt	photometric
	silver diethyldithiocarbamate	photometric
	Thionalid	gravimetric
Arsenic	Eriochrome Cyanin	photometric
	potassium xanthate	photometric
	silver diethyldithiocarbamate	photometric
	thioacetamide	gravimetric
	Thionalid	photometric
	thiourea	gravimetric
Beryllium	Alberon	photometric
	Berillon II IREA	photometric
	Curcumin	photometric
	4-p-nitrophenylazo-orcinol	photometric
	quinalizarin	photometric
	tannin	gravimetric
Bismuth	Cupferron	gravimetric
	diantipyrilmethane	gravimetric
	Dithizone	volumetric
	8-hydroxyquinoline	gravimetr., volumetr., photometr.
	mercaptobenzothioazole	gravimetric

Table 28 (continued)

Element	Reagent	Method
Bismuth (cont.)	mercaptophenylthiodiazolone	photometric
	picric acid	gravimetric
	pyrogallol	gravimetric
	sodium diethyldithiocarbamate	photometr., solv., extr.
	thioacetamide	gravimetric
	Thionalid	photometric
	thiourea	gravimetric
	Thoron	photometric
	Xylenol Orange	photometric
Cadmium	anthranilic acid	gravimetric
	Cadion	photometric
	Cadion IREA	photometric
	Crystal Violet	photometric
	diantipyrylmethane	gravimetric
	diphenylcarbazide	photometric
	Dithizone	photometric
	8-hydroxyquinoline	gravimetr., volumetr., photometr.
	mercaptobenzothiazolone	gravimetric
	Methyl Violet	photometric
	α-naphthoquinone	gravimetric
	quinaldic acid	gravimetric
	sodium diethyldithiocarbamate	heterometric
	tetraphenylarsonium chloride	photometric
	thioacetamide	gravimetric
	thiourea	gravimetric
Calcium	Azoazoxy	photometric
	chloranilic acid	photometric
	oxalic acid	gravimetric
	picrolonic acid	gravimetric
	sodium naphthalenehydroxamate	photometric
Cerium	8-hydroxyquinoline	extr. + photometric
	1-thenoyltrifluoroacetone	extr. + photometric
Cesium	dipicrylamine	gravimetr., photometr.
	Kalignost	gravimetr., volumetr., photometr.
Chromium(III)	8-hydroxyquinoline	gravimetr., volumetr.
	Komplexon III	photometric
	quinaldic acid	gravimetric
Chromium(VI)	chromotropic acid sodium salt	photometric
	diphenylcarbazide	photometic
	Komplexon III	photometric

TABLE 28 (continued)

Element	Reagent	Method
Cobalt	anthranilic acid	gravimetric
	diantipyrylmethane	gravimetric
	dimethylglyoxime	photometric
	Dithizone	photometric
	8-hydroxyquinoline	gravimetr., volumetr., photometr.
	8-mercaptoquinoline sodium salt	photometric
	N,N'-ethylene-di-(4-methoxy-1,2-benzo-quinone-1-oxime-2-imine)α-nitroso-β-naphthol	gravimetr., photometr.
	Nitroso-R salt	photometric
	PAN	photometric
	phenyldimethylpyrazolone	photometric
	quinaldic acid	gravimetric
	rubeanic acid	gravimetric
	thenoyltrifluoroacetone	extr. + photometric
	thiourea	gravimetric
Copper	anthranilic acid	gravimetric
	α-benzoinoxime	gravimetric
	2,2'-bicinchonin	photometric
	Cupferron	gravimetric
	diallyldithiocarbamoylhydrazine	gravimetric
	1,2-diaminoanthraquinone-3-sulfonic acid	photometric
	p-dimethylaminobenzylidene-rhodanine	photometric
	2,9-dimethyl-4,7-diphenyl-1,10-phenanthroline	photometric
	dimethylglyoxime	photometric
	1,3-dimethyl-4-imino-5-oxyimino-alloxane (Daxime)	photometric
	2,2'-diquinolyl (Cuproine)	extr. + photometric
	8,8-diquinolyl disulfide	photometric
	Dithizone	extr. + photometric
	8-hydroxyquinoline	gravimetr., volumetr., photometr.
	mercaptoacetic acid	photometric
	mercaptobenzothiazole	gravimetr. + volumetr.
	Neocuproine	extr. + photometr.
	α-nitroso-β-naphthol	photometric
	potassium xanthate	photometric
	quinaldic acid	gravimetric
	salicylaldoxime	gravimetric
	salicylic acid	photometric
	sodium diethyldithiocarbamate	extr. + photometr.
	thenoyltrifluoroacetone	extr. + photometr.
	thioacetamide	gravimetric
	Thionalid	gravimetric
	thiosemicarbazide	photometric
	thiourea	gravimetric
Gallium	alizarinsulfonic acid sodium salt	photometric
	Cupferron	gravimetric

TABLE 28 (continued)

Element	Reagent	Method
Gallium (cont.)	dibromooxyquinoline	gravimetric
	Gallion IREA	photometric
	8-hydroxyquinoline	gravimetr., volumetr., photometr.
	morin	photometric
	quinalizarin	extr. + photometric
	Rhodamine B	extr. + photometric
	tannin	gravimetric
Germanium	diphenylcarbazone	photometric
	8-hydroxyquinoline	gravimetric
	phenylfluoron	photometric
Gold	p-dimethylaminobenzylidene-rhodanine	photometric
	Dithizone	photometric
	mercaptobenzothiazole	gravimetric
	Rhodamine B	photometric
	thiophenol	gravimetric
	thiourea	gravimetric
	o-toluidine	photometric
Hafnium	Arsenazo III	photometric
	Cupferron	gravimetric, volumetric
Indium	Arsenazo	photometric
	5,7-dibromo-8-hydroxyquinoline	photometric
	Dithizone	photometric
	8-hydroxyquinoline	gravimetr., volumetr., photometr.
	Rhodamine 6G	photometric
	thiooxin	photometric
Iron(II)	cyclohexanedionedioxime	photometric
	dimethylglyoxime	photometric
	4,7-diphenyl-1,10-phenanthroline	extr. + photometr.
	2,2'-dipyridyl	photometric
	mercaptoacetic acid	photometric
	1,10-phenanthroline	photometric
	quinaldic acid	gravimetric
	sulfosalicylic acid sodium salt	photometric
Iron(III)	acetylacetone	extr. + photometr.
	Cupferron	gravimetric
	diantipyrylmethane	gravimetric
	8-hydroxyquinoline	gravimetr., volumetr., photometr.
	mercaptoacetic acid	photometric
	α-nitroso-β-naphthol	photometric
	quinaldic acid	gravimetric
	salicylaldoxime	photometric
	salicylic acid	photometric

TABLE 28 (continued)

Element	Reagent	Method
Iron(III) (cont.)	sulfosalicylic acid sodium salt Tiron	photometric photometric
Lanthanides, Y and Sc	Arsenazo I Arsenazo III Thoron	photometric photometric photometric
Lead	anthranilic acid Arsacene diallyldithiocarbamoylhydrazine diphenylcarbazide Dithizone 8-hydroxyquinoline mercaptobenzothiazole phthalic acid picrolonic acid quinaldic acid salicylaldoxime sodium diethyldithiocarbamate Sulfarsacene thioacetamide Thionalid thiourea	gravimetric photometric gravimetric photometric photometric gravimetr., volumetr., photometr. gravimetric photometric gravimetric gravimetric gravimetric photometric volumetric gravimetric gravimetric gravimetric
Lithium	8-hydroxyquinoline Thoron	gravimetr., volumetr., photometr. photometric
Magnesium	bisalicylideneethylenediamine diphenylcarbazide 8-hydroxyquinoline Magneson IREA phenazone picrolonic acid quinalizarin sulfanilic acid Titan Green	photometric photometric gravimetr., volumetr., photometr. photometric photometric gravimetric photometric photometric photometric
Manganese	anthranilic acid cystein 8-hydroxyquinoline 8-mercaptoquinoline sodium salt Nioxime quinaldic acid sodium diethyldithiocarbamate thiourea	gravimetric gravimetric gravimetr., volumetr., photometr. photometric photometric gravimetric photometric gravimetric
Mercury	anthranilic acid Crystal Violet	gravimetric photometric

Table 28 (continued)

Element	Reagent	Method
Mercury (cont.)	p-dimethylaminobenzylidene-rhodanine	photometric
	diphenylcarbazide	photometric
	diphenylcarbazone	photometric
	Dithizone	photometric
	Methyl Violet	photometric
	tetraphenylarsonium chloride	photometric
	thioacetamide	gravimetric
	Thionalid	gravimetric
	thiourea	gravimetric
Molybdenum	α-benzoinoxime	gravimetric
	Dithiol	photometric
	8-hydroxyquinoline	gravimetr., volumetr., photometr.
	8-mercaptoquinoline sodium salt	photometric
	phenylhydrazine	extr. + photometr.
	tannin	gravimetric
	thioacetamide	gravimetric
Nickel	anthranilic acid	gravimetric
	α-benzyldioxime	gravimetric
	diallyldithiocarbamoylhydrazine	gravimetric
	dimethylglyoxime	gravimetr., volumetr., photometr.
	Dithizone	photometric
	α-furildioxime	gravimetr., volumetr.
	8-hydroxyquinoline	gravimetr., photometr.
	Nioxime	photometric
	quinaldinic acid	gravimetric
	sodium diethyldithiocarbamate	extr. + photometr.
	thiourea	gravimetric
Niobium	Cupferron	gravimetric
	phenylarsonic acid	gravimetric
	pyrogallol	photometric
	tannin	gravimetric
	Thoron	photometric
Osmium	8-mercaptoquinoline sodium salt	photometric
	Thionalid	photometric
	thiourea	photometric
Palladium	acetylene	photometric
	p-dimethylaminobenzylidene-rhodanine	photometric
	dimethylglyoxime	gravimetr., volumetr., photometr.
	2,2'-dipyridyl	gravimetric
	Dithizone	extr. + photometric
	α-furildioxime	gravimetr. extr. + photometr.
	8-hydroxyquinoline	gravimetr., volumetr., photometr.
	8-mercaptoquinoline sodium salt	photometric
	α-nitroso-β-naphthol	photometric
	1,10-phenanthroline	photometric
	phenylpyridyl ketoxime	photometric
	salicylaldoxime	gravimetric
	thiourea	gravimetric

TABLE 28 *(continued)*

Element	Reagent	Method
Platinum	α-furildioxime 2-mercaptobenzothiazole	gravimetric gravimetric
Potassium	dipicrylamine Nitroso-R salt sodium tetraphenylboron	gravimetr., volumetr., photometr. photometric gravimetr., volumetr.
Rhenium	8-mercaptoquinoline sodium salt Nitron Rhodamine 6G	photometric gravimetric photometric
Rhodium	thiobarbituric acid Thionalid thiourea	gravimetric gravimetric gravimetric
Rubidium	dipicrylamine sodium tetraphenylboron	gravimetr., photometr. photometr., volumetr.
Ruthenium	anthranilic acid 8-mercaptoquinoline sodium salt Thionalid thiourea	photometric photometric gravimetric gravimetric
Scandium	Arsenazo III 8-hydroxyquinoline sulfonazo	photometric gravimetr., photometr. photometric
Silver	diallyldithiocarbamoylhydrazide p-dimethylaminobenzylidene-rhodanine diphenylcarbazone Dithizone mercaptobenzothiazole quinaldic acid sodium diethyldithiocarbamate thiourea	gravimetric volumetr., photometr. photometric photometric gravimetric volumetric extr. + photometr. gravimetric
Strontium	chloranilic acid picrolonic acid	photometric gravimetric
Tantalum	Cupferron p-dimethylaminobenzylidene-rhodanine phenylarsonic acid phenylfluoron pyrogallol	gravimetric photometric gravimetric photometric photometric

SUMMARIZING TABLES

TABLE 28 (continued)

Element	Reagent	Method
Tantalum (cont.)	tannin Thoron	gravimetric photometric
Thallium	Brilliant Yellow Crystal Violet Dithizone mercaptobenzothiazole 8-mercaptoquinoline sodium salt Methyl Violet Rhodamine B Rhodamine 6G sodium tetraphenylboron Thionalid thiourea	photometric photometric photometric gravimetric photometric photometric photometric photometric gravimetr., volumetr., photometr. gravimetric gravimetric
Thorium	alizarin alizarinsulfonic acid sodium salt Arsenazo III Cupferron 8-hydroxyquinoline morin picrolonic acid Thoron	photometric photometric photometric gravimetric gravimetric photometric gravimetric photometric
Tin	Brilliant Yellow Crystal Violet Dithiol Dithizone 8-hydroxyquinoline mercaptoacetic acid Methyl Violet phenylfluoron pyrogallol Rhodamine B thioacetamide thiourea 2,3,7-trihydroxy-9-*p*-nitrophenyl-6-fluoron	photometric photometric photometric photometric gravimetr., volumetr., photometr. photometric photometric photometric gravimetric photometric gravimetric gravimetric photometric
Titanium	Cupferron 8-hydroxyquinoline sulfosalicylic acid Tiron	gravimetr., photometr. extr. + photometr. photometric photometric
Tungsten	dithiol 8-hydroxyquinoline mercaptoacetic acid Rhodamine B	photometric gravimetr., volumetr., photometr. photometric photometric

TABLE 28 (continued)

Element	Reagent	Method
Uranium	Arsenazo III	photometric
	Cupferron	gravimetric
	mercaptoacetic acid	photometric
	morin	photometric
	PAN	photometric
	quinaldic acid	gravimetric
	thenoyltrifluoroacetone	photometr., extr.
	Thoron	photometric
Vanadium	Cupferron	gravimetric
	8-hydroxyquinoline	gravimetr., volumetr., photometr.
	8-mercaptoquinoline sodium salt	photometric
	PAN	extr. + photometr.
Zinc	anthranilic acid	gravimetric
	Arsacen	photometric
	Brilliant Yellow	photometric
	diallyldithiocarbamoylhydrazone	gravimetric
	Dithizone	extr. + photometr.
	8-hydroxyquinoline	gravimetr., volumetr., photometr.
	Methyl Violet	photometric
	quinaldic acid	gravimetric
	Rhodamine B	photometric
	sodium diethyldithiocarbamate	photometric
	Sulfarzacene	gravimetr., volumetr., photometr.
	Thoron	photometric
	8-tosylaminoquinoline	photometric
	Xylenol Orange	photometric
Zirconium	Alizarin	photometric
	alizarinsulfonic acid sodium salt	photometric
	Arsenazo	photometric
	Arsenazo III	photometric
	chloranilic acid	photometric
	Cupferron	gravimetric
	Dithizone	photometric
	8-hydroxyquinoline	gravimetr., volumetr., photometr.
	mandelic acid	gravimetric
	morin	photometric
	phenylarsonic acid	gravimetric
	purpurin	photometric
	Pyrocatechol Violet	photometric
	tannin	gravimetric
	Thoron	photometric
	Zirconon	gravimetric

SPECTROPHOTOMETRIC DETERMINATION OF METALS

Efforts were made in the compilation of Table 29 to enable the experienced analyst not only to find the details with regard to reagents and disturbing ions involved in the photometric determination of a given metal ion, but even to accomplish the analysis, in the majority of cases, without consulting the original reference. However, at least one reference is still provided in each case to facilitate the gaining of further information.

The first column of the Table contains the metal ions in alphabetical order. The reagents useful in the determination of the given metal are listed in the second column, again in alphabetical order. The third column contains the wavelengths of the absorption maxima of the complexes at which the absorption of the ligand is minimum and the interfering effects of other accompanying constituents and metal ions are also minimum. The fourth column indicates that concentration region within which the absorption of the complex follows the Beer–Lambert Law. The fifth column specifies the most important reaction conditions, mainly the pH of the solution, the masking agents, the organic solvent in the case of solvent extraction procedures, etc. The composition of the buffer solution used for the adjustment of the reaction medium is given in only those cases where it has particular significance. The sixth column lists the interfering ions. There are methods in which the interfering ions have not yet been specified. In such cases, as well as for quite specific methods, this column is left empty. The last column contains the references. In order to save space, the authors' names have been omitted. As far as possible, the more recent sources have been included which contain references to the earlier ones.

TABLE 29. METAL DETERMINATION BY SPECTROPHOTOMETRY

Metal ion	Reagent	Absorption maximum, nm	Concentration range, µg/ml	Reaction conditions	Interfering ions†	Refs.
Ag^+	p-Diethylaminobenzylidene-rhodanine	470–530	0.004–0.04 0.1–2.0	0.05 N strong mineral acid	Pd^{2+}, Au^{III} Hg^{2+}, Cu^{2+}	Anal. Chim. Acta **5**, 445 (1951); **9**, 80 (1953)
	Dithizone	460	0.2–2.0	0.5 N strong mineral acid; extr. CCl_4 $CHCl_3$	Pd^{2+}, Au^{III}, Hg^{2+}	Anal. Chem. **24**, 1503 (1952); **27**, 305 (1955) Z. Anal. Chem. **101**, 11 (1935)
Al^{3+}	alizarinsulfonic acid sodium salt	485	0.1–1.2	pH 4.5	Be^{2+}, Bi^{3+}, Cu^{2+}, Fe^{3+}, Mo^{IV}, V^V, W^{VI}	Anal. Chim. Acta **4**, 517 (1950)
	Aluminon	525	0.04–0.4	pH 4.7–5.5	Ag^+, Be^{2+}, Cr^{3+}, Cu^{2+}, Pb^{2+}, Sc^{3+}, Sn^{2+}, Th^{4+}, Ti^{4+}, V^V, Zr^{IV}	I. E. C. Anal. Ed. **17**, 206 (1945) Anal. Chim. **24**, 1120, 1122 (1952) Zav. Lab. **20**, 414 (1954)
	Eriochrome Cyanin	535	0.04–0.4	pH 4–6	Be^{2+}, V^V	Anal. Chim. **28**, 1419 (1956)
		587				Bunseki Kagaku **18**, 323 (1969) Z. Chem. **7**, 438 (1967)
	8-hydroxyquinoline	390	2–120	pH 5.0–9.5; extr. $CHCl_3$		Anal. Chim. Acta **10**, 373 (1954)

	Reagent	λ	Range	Conditions	Interferences	Reference
Au^{III}	p-dimethylaminobenzylidene-rhodanine	500	0.1–1.5	0.1 N HCl	Pd^{2+}, Pt^{IV}, Ag^+, Hg^{2+}, Fe^{3+}, Cu^{2+}, Pb^{2+}	Anal. Chem. **20**, 253 (1948)
	Rhodamine B	545	0.3–3	extr. with isopropyl ether		Anal. Chim. Acta **13**, 154 (1955)
	o-toluidine	437	0.04–0.4	0.5 N H_2SO_4	OsO_4, Ru^{IV}, V^V, halogens	J. N. B. S. **36**, 119 (1946)
Be^{2+}	acetylacetone	295	>0.25	extr.		Anal. Chim. Acta **6**, 462 (1952)
	Aluminon	506, 535	0.02–4.0	pH ≈ 5.0	Al^{3+}, Ca^{2+}, Cu^{2+}, Fe^{2+}, Mn^{2+}, Pb^{2+}	Anal. Chem. **22**, 936 (1950)
	Berillon II	600	0.04–0.45	pH 12–13.2	Ca^{2+}, Mg^{2+}, Al^{3+}, Fe^{3+}, Cu^{2+}, Ni^{2+}	Zh. Anal. Khim. **11**, 393, 400 (1956); Zav. Lab. **23**, 280 (1957)
	Eriochrome Cyanin	512	0.04–0.4	pH 9.8	Co^{2+}, Mn^{2+}, Cr^{3+}	Anal. Chem. **30**, 521 (1958)
	8-hydroxyquinoline	320–370	0–60	pH 9.2; extr. methyl isobutyl ketone		Sci. Rap. Tokuku Univ. **A12**, 334 (1960)
	4-(p-nitrophenylazo)-orcinol	515	1–8	0.1 M NaOH + EDTA	Ca^{2+}, Cu^{2+}, Fe^{2+}, Mg^{2+}, Ni^{2+}, Zn^{2+}	Anal. Chem. **25**, 1583 (1953); **28**, 1728 (1956)
	quinalizarin	650	0.1–10.0	pH >10.0	Co^{2+}, Fe^{2+}, Mg^{2+}, Ni^{2+}, Sc^{3+}, Ti^{4+}, Zr^{IV}	Z. Angew. Chem. **160**, 729 (1948)

† Note: Roman numbers in this column refer to the formal oxidation state of the atom.

TABLE 29 (continued)

Metal ion	Reagent	Absorption maximum, nm	Concentration range, μg/ml	Reaction conditions	Interfering ions‡	Refs.
Be^{2+} (cont.)	sulfosalicylic acid	317	0.1–1	0.5 M EDTA; pH 9.2–10.5	Al^{3+}, Ce^{4+}, Cu^{2+}, Fe^{3+}, NO_3^-, PO_4^-	Anal. Chem. **22**, 1512 (1950)
Bi^{3+}	Dithizone	490	>1.0	pH 3.0–9.0; extr. CCl_4	Pb^{2+}, Tl^I	Zh. Anal. Khim. **8**, 286 (1953)
	Komplexon III	263.5	2–25	pH 0.8–1.2	Fe^{3+}, Cu^{2+}	Acta Chim. Hung. **28**, 143 (1961) Nippon Kinzoku Gakkaislu **31**, 893 (1967) Anal. Chem. **27**, 1221 (1955)
	sodium diethyldithio-carbamate	360–370	1–30	pH 8.5–11 + Komplexon III and acetone	Cu^{2+}, Fe^{3+}, Hg^{2+}, Sb^{3+}, Sn^{3+}	Anal. Chem. **27**, 24 (1955)
	thiourea	322	3–30	0.2–0.4 N H_2SO_4	Os^{VIII}, Ru^{IV}, Ag^+, Fe^{3+}, Sb^{3+}, Hg^{2+}, Pb^{2+}, Cu^{2+}, Cd^{2+}, Sn^{2+}	Anal. Chem. **27**, 1722 (1955)
		420	5–25.0	0.4–1.2 N HNO_3		Z. Anal. Chem. **97**, 96 (1934)
Ca^{2+}	Murexide	506	0.1–1.2	pH 11.3	Ba^{2+}, Fe^{3+}, Hf^{IV}, Hg^{2+}, Li^+, Mg^{2+}, Sr^{2+}, Zr^{IV}, SO_4^{2-}	Anal. Chem. **25**, 1414 (1953) Analyst **81**, 350 (1956)

SUMMARIZING TABLES

Cd^{2+}	Cadion	560	0.05–0.5		Fe^{3+}, Hg^{2+}, Mg^{2+}, Ni^{2+}, Sn^{2+}, CN^-	Anal. Chim. Acta **19**, 377 (1959)
	Dithizone	518	0.1–1.3	pH 4–12; extr. CCl_4	Hg^{2+}, Tl^{3+}, Ag^+, Co^{2+}, Cu^{2+}	Anal. Chem. **25**, 493 (1953) Zh. Anal. Khim. **23**, 1847 (1968)
	4-hydroxy-3-nitrophenyl-arsonic acid	410	3–16			Chem. Zt. **79**, 364 (1955)
Ce^{3+}	benzidine	315	0.4–5	HCl	V^V, Cr^{VI}	J. Soc. Chem. Ind. **51**, 647 (1932)
	brucine	400–430	>10.0	1 N H_2SO_4	Mn^{2+}, Cr^{VI}, Fe^{3+}, $S_2O_8^{2-}$	Zh. obshch. Khim. **9**, 698 (1939)
	8-hydroxyquinoline	505	2–20	pH 9.9–10.5	V^V, F^-	Analyst **73**, 275 (1948)
Co^{2+}	dimethylglyoxime + I^-	540	250–3750	pH 6		Acta Chim. Hung. **45**, 77 (1965)
	Komplexon III + H_2O_2			pH > 5		Chem. Listy. **45**, 237 (1951) Anal. Chim. Acta **6**, 278 (1952)
	N,N'-ethylene-di(4-methoxy-1,2-benzoquinone)-1-oxime-2-imine	381	0.02–20	pH 2–7; extr. $CHCl_3$		J. Chem. Soc. Japan **80**, 1260, 1263 (1959) Japan Analyst **9**, 269, 272 (1960)
	nitrosocresol	360	0.02–25	extr. petr. ether		I. E. C. Anal. Ed. **17**, 254 (1945)
	1-nitroso-2-naphthol	365 317	>0.05 0.2–10	pH 8 pH 4.0–5.5	Cu^{2+}, Fe^{2+}	Anal. Chem. **27**, 1731 (1955)

† Note: Roman numbers in this column refer to the formal oxidation state of the atom.

TABLE 29 (continued)

Metal ion	Reagent	Absorption maximum, nm	Concentration range, μg/ml	Reaction conditions	Interfering ions‡	Refs.
Co^{2+} (cont.)	2-nitroso-1-naphthol	530	0.2–4	pH 3–4; extr. $CHCl_3$	$Fe^{2+}, Pd^{2+}, Sn^{2+}$	Anal. Chim. Acta **12**, 547 (1955)
	sodium diethyldithio-carbamate	367	2–25	pH 5–11; extr. CCl_4	Bi^{3+}, CN^-	Z. Anal. Chem. **143**, 182 (1954)
	2,2′,2″-terpyridyl	482 505	0.5–25	pH 2.0–10.0	Cr^{3+}, Fe^{3+}, V^V	I. E. C. Anal. Ed. **15**, 74 (1943); Anal. Chem. **26**, 1968 (1954)
Co^{III}	Komplexon III	540	1.0–11.0	pH 7.0–8.0	$Cu^{2+}, Fe^{3+}, Cr^{3+}$	Chem. Listy **45**, 237 (1951)
	Nitroso-R salt	425	0.1–5.0	pH 5.0–6.5	$Ni^{2+}, Fe^{3+}, Cu^{2+}, V^V, W^{VI}, Mo^{VI}$	Z. Anal. Chem. **144**, 165 (1956); Anal. Chem. **28**, 1151 (1956)
	sodium diethyldithio-carbamate	367 650	0–6	pH 4–11; extr. CCl_4		Anal. Chim. Acta **6**, 278 (1952); Chem. Listy **46**, 603 (1952)
Cr^{3+}	Komplexon III	550	1–80	pH 3.0–5.0 boiling		Collection Czech. Chem. Comm. **15**, 42 (1950)
Cr^{VI}	diphenylcarbazide	540	0.01–2.6	0.2–2.0 N H_2SO_4	$Mo^{VI}, Hg^{2+}, Fe^{3+}, V^V$	Anal. Chem. **22**, 1317, (1950); **30**, 359, 447 (1958); Z. Anal. Chem. **233**, 415 (1968)

	Reagent				Reference
Cu^{2+}	2,9-dimethyl-1,10-phenanthroline (Neocuproine)	454	1–10	pH 3–9; extr. with propanol or isoamyl alcohol		*Anal. Chem.* **24**, 371 (1952)
	2,9-dimethyl-4,7-diphenyl-1,10-phenanthroline (Bathocuproine)	479	0.3–5.0	pH 5–7 extr. propanol		*Anal. Chem.* **25**, 510 (1953)
	1,3-dimethyl-4-imino-5-oxyiminoalloxane (Daxime)	382	1–13	pH 7–9.5		*Talanta* **8**, 77 (1961)
	2,2'-diquinolyle (Cuproine)	545	0.5–4	pH 4.4–7.5; extr. amyl alcohol	Co^{2+}, Pd^{2+}, Ni^{2+}, Fe^{2+}	*Anal. Chem.* **25**, 1484 (1953) *Anal. Chim. Acta* **9**, 263 (1953)
	Dithizone	545	0.01–1.0	pH 2–3; extr. CCl_4		*Anal. Chim. Acta* **8**, 197 (1953)
	Komplexon III	700–750 700–750	35–3600 80–8000	pH 4.75–6.50 pH 10.0	Ag^+, Au^{III}, Bi^{3+}, Fe^{3+}, Hg^{2+}, Pd^{2+}	*Z. Anal. Chem.* **143**, 1 (1954)
	nitrilotriacetic acid (Komplexon I)	690–710	40–7000	pH 4.0–5.0		*Z. Anal. Chem.* **144**, 401 (1953)
	pyridine + SCN^-	405	3–10	pH 5–7; extr. $CHCl_3$		*Ann. Chim. Appl.* **16**, 96 (1926)
	pyridine + SCN^-	650	>1.0	extr. CCl_4	Fe^{3+}, Ni^{2+}, Co^{2+}, Zn^{2+}	*Analyst* **55**, 187 (1930) *Anal. Chem.* **22**, 612 (1950) *Zh. Priklad. Khim.* **17**, 252 (1944)
	sodium diethyldithiocarbamate	436	0.1–5.0	pH 4–11; extr. CCl_4	Tl^{III}, Bi^{3+}, Au^{III}	*Z. Anal. Chem.* **143**, 182 (1954)

† Note: Roman numbers in this column refer to the formal oxidation state of the atom.

TABLE 29 *(continued)*

Metal ion	Reagent	Absorption maximum, nm	Concentration range, µg/ml	Reaction conditions	Interfering ions†	Refs.
Cu^{2+} *(cont.)*	zinc dibenzyldithio-carbamate	435	0.5–4	1 M HCl	Ag^+, Bi^{3+}, Co^{2+}, Fe^{3+}, Hg^{2+}, Ni^{2+}, Sb^{3+}	*Anal. Chem.* **24**, 991 (1952)
Fe^{2+}	dimethylglyoxime	500–600	>2.0	pH ≈ 8–9; pyridine	Al^{3+}, Mg^{2+}, Zn^{2+}	*Anal. Chem.* **19**, 1017 (1947); *J. Chem. Soc. Japan* **75**, 1069 (1954)
	2,2′-dipyridyl	552	0.01–1.5	pH 3.0–9.0	Ag^+, Bi^{3+}, Cd^{2+}, Zn^{2+}, Be^{2+}, Mo^{VI}, W^{VI}, Cu^{2+}, Sn^{2+}	*I. E. C. Anal. Ed.* **14**, 862 (1942)
	1,10-phenanthroline	510	0.5–5	pH 2.0–9.0	Ag^+, Bi^{3+}, Hg^{2+}, Mo^{VI}, Ni^{2+}, W^{VI}	*I. E. C. Anal. Ed.* **10**, 60 (1938)
	2,2′,2″-tripyridyl	552	0.5–5	pH 3.0–10.0	Co^{2+}, Cu^{2+}, Ni^{2+}, W^{VI}	*I. E. C. Anal. Ed.* **14**, 862 (1942)
Fe^{3+}	4,7-diphenyl-1,10-phenanthroline	535	0.2–2.5	pH 4–5; extr. with isoamyl alcohol		*Anal. Chem.* **25**, 1337 (1953); **28**, 809 (1956)
	Ferron	610	>0.015	pH 2.7–3.1		*I. E. C. Anal. Ed.* **9**, 406 (1937)

SUMMARIZING TABLES

Ion	Reagent	λ	Range	Conditions	Interferences	Reference
Fe^{3+} (cont.)	Komplexon III + H_2O_2	530	0–70	pH 9.0–10.0	CN^-	*Chem. Listy* **50**, 899 (1956)
	Komplexon III	260	0.006–17	0.1 M HCl	Cu^{2+}, Hg^{2+}, Tl^{4+}, NO_3^-	*J. Chem. Soc. Japan* **28**, 88 (1955)
	salicylaldoxime	480	>0.01	pH 6.2–6.6	many elements	*I.E.C. Anal. Ed.* **12**, 448 (1940)
	salicylic acid	520	0.03	pH 2.5–2.7	many elements	*Analyst* **66**, 142 (1941)
	sulfosalicylic acid	450–500	>0.04	pH 4.5–7.0	Cu^{2+}, Ni^{2+}, Co^{2+}, Cr^{3+},	*Chem. Weekblad.* **42**, 311 (1946)
	tartaric acid	424	0–5.4	pH 8.5–11.5	Ni^{2+}, Cu^{2+}, Co^{2+}	
		366	1.2–24	pH 1.5–2.3 pH 3.5–7.1		*Metall.* **8**, 374 (1954)
Ga^{3+}	Rhodamine B	365	0.05–1.0	6 M HCl; extr. with diethyl ether		*Anal. Chim. Acta* **13**, 159 (1955)
		530 565	0.1–1.0	6 M HCl; extr. with benzene + diethyl ether	Sb^{3+}, Au^{3+}, Fe^{3+}, Tl^{3+}, W^{VI}	
	quinalizarin		>0.02	pH 5.0	many elements	*J. Amer. Chem. Soc.* **59**, 40 (1937)
Ge^{VI}	phenylfluoron	490–510	0.02–1.0	0.6–1.8 N HCl	Sb^{3+}, Sn^{4+}, As^{3+}, PO_4^{3-}, Ce^{4+}	*Anal. Chem.* **28**, 1273, (1956)
		508			Mo^{VI}, Nb^V	*Bunseki Kagaku* **16**, 715 (1967)†

† Note: Roman numbers in this column refer to the formal oxidation state of the atom.

TABLE 29 (continued)

Metal ion	Reagent	Absorption maximum, nm	Concentration range, µg/ml	Reaction conditions	Interfering ions†	Refs.
Hg^{2+}	diphenylcarbazone		0.1	pH 5.7; extr.	many elements	J. Chem. Soc. Japan **82**, 452 (1961)
	Dithizone	490 625	0.1–3.0	pH 1.0–1.5; extr. $CHCl_3 + CCl_4$	Ag^+, Cu^{2+}, Pd^{2+}, Pt^{IV}, Au^{III}, Bi^{3+}, Cl^-, Br^-	Analyst **72**, 6 (1947) U.S. At. Energy Comm. Rept. IN-1159, p. 22 (1968)
In^{3+}	5,7-dibromo-8-hydroxyquinoline	415	0.2–20	pH 3.5–4.5		Anal. Chem. **30**, 2055 (1958)
	8-hydroxyquinoline	400	0.1–20.0	pH 3.2–4.5; extr. $CHCl_3$	Al^{3+}, Ga^{3+}, Fe^{3+}, Sn^{IV}, Bi^{3+}, Cu^{2+}, Tl^{3+}, V^V, Mo^{VI}, Ni^{2+}	Analyst **82**, 549 (1957); Anal. Chem. **28**, 1340 (1956)
Ir^{IV}	Komplexon III	313	5–60	pH 11.4–12.6	Pt-elements, NO_3^-	Anal. Chem. **28**, 16 (1956)
K^+	dipicrylamine + Li_2CO_3	450	0.2–1	pH > 8	Ba^{2+}	Analyst **80**, 768 (1955)
	sodium tetraphenylboron	266	2–30	dissol. of ppt. in 75% CH_3CN	Ag^+, Co^+, Hg^{2+}, NH_4^+, Rb^+, Tl^+	Anal. Chem. **28**, 1542 (1956)
Li^+	Thoron	460–500	<10.0	1% KOH	Ca^{2+}, Mg^{2+}	Anal. Chem. **28**, 1527 (1956)

SUMMARIZING TABLES

Mg^{2+}	8-hydroxyquinoline	388	<1.0	pH ≈ 9–11.5; extr. $CHCl_3$		*Anal. Chim. Acta* **17**, 234 (1957)
		357	0–60	pH 10.2; extr. methyl isobutyl ketone		*Sci. Rep. Tokuku Univ.* **A12**, 334 (1960)
	Eriochrome Black T	520	0.2–1.4	pH 8–10	Al^{3+}, Ca^{2+}, Co^{2+}, Fe^{2+}, Mn^{2+}, Ni^{2+}	*Anal. Chem.* **25**, 498 (1953) *Analyst* **81**, 348 (1956)
	Titan Yellow	535	0.1–1.5	pH > 12	Al^{3+}, Cd^{2+}, Co^{2+}, Ni^{2+}, Zn^{2+}, Sn^{2+}, Cu^{2+}, As^{3+}, Hg^{2+}, NH_4^+, Ca^{2+}, Ba^{2+}, Sr^{2+}	*Zh. Anal. Khim.* **22**, 1423 (1967) *Z. Anal. Chem.* **136**, 254 (1952) *Analyst* **75**, 81 (1950)
Mn^{2+}	Komplexon III	500	1–10		Sn^{2+}, Mn^{2+}, Fe^{2+}, Cr^{3+}	*Collection Czech. Chem. Comm.* **15**, 456 (1950)
Mo^V	mercaptoacetic acid	352.5	1–40	pH 0.7–7		*Pharm. Zhalle.* **89**, 3 (1950); *Anal. Chem.* **30**, 1282 (1958)
	thiomalic acid	352.5	1–40	0.5 N HCl		
Mo^{VI}	4-methyl-1,2-dimercapto-benzene (dithiol)	670	0.05–1	1.5 M H_2SO_4		*Analyst* **72**, 185 (1947) *Z. Anal. Chem.* **151**, 129 (1956)
	mercaptoacetic acid	365	1–40	pH 3–6		*Pharm. Zhalle.* **89**, 3 (1950); *Anal. Chem.* **30**, 1282 (1958)
	thiomalic acid	365	1–40	pH 3–6		

† Note: Roman numbers in this column refer to the formal oxidation state of the atom.

TABLE 29 (continued)

Metal ion	Reagent	Absorption maximum, nm	Concentration range, μg/ml	Reaction conditions	Interfering ions†	Refs.
Nb^V	8-hydroxyquinoline	385	1–8	pH ≈ 9.4; extr. $CHCl_3$		Anal. Chem. **27**, 492 (1955)
	pyrogallol		>0.3	pH 6.0–8.0	Ti^{4+}	Chem. Zentr. **2**, 2211 (1937)
Ni^{2+}	α-benzyldioxime	275 406	0–1.0	extr. $CHCl_3$		Zh. Anal. Khim. **16**, 596 (1961)
	dimethylglyoxime + bromine water	520–540	0.1–1.2	pH > 3.0–5	Fe^{3+}, Al^{3+}, Mg^{2+}, Co^{2+}, Cu^{2+}, Mn^{2+}	I. E. C. Anal. Ed. **17**, 380 (1945)
	dimethylglyoxime	366	1–15	pH 6.5–8.5; extr. $CHCl_3$		Z. Anal. Chem. **140**, 267 (1953); **143**, 272 (1954); **150**, 114 (1956); Anal. Chem. **32**, 522 (1962) Zavod. Lab. **33**, 921 (1967)
	Komplexon III	580–720	0.4–50	pH 5.5–6.8		Anal. Chim. Acta **11**, 367 (1954)
	β-mercaptopropionic acid	330	0.5–10	pH 8.5–9.5	Fe^{3+}	Anal. Chem. **25**, 1411 (1953)
	sodium diethyldithio-carbamate	328	2–25	pH 5–11; extr. CCl_4		Z. Anal. Chem. **143**, 182 (1954)
Os^{VIII}	thiourea	450–530	0–3 8–40	4 N HCl; 0.6 N H_2SO_4	Ru^{IV}, Pd^{2+}	Anal. Chem. **22**, 317 (1950)
Pb^{2+}	Dithizone	510–530	0.05–0.4	pH 8–10; extr. CCl_4 pH 8.5–11; extr. $CHCl_3$	Sn^{2+}, Bi^{3+}, Pt^{IV}, Pd^{2+}, Au^{III}, Ag^+, Hg^{2+}, Zn^{2+}, Cd^{2+}	Anal. Chem. **30**, 1139 (1958) Anal. Chem. **26**, 1747 (1954)

	Reagent	λ (nm)	Range	Conditions	Interferences	Reference
Pb^{2+} (cont.)	1-nitroso-2-naphthol	370	5–25	pH 1.0–2.0; extr. toluene		Z. Anal. Chem. **148**, 283 (1955)
	p-nitrosodiphenylamine	525	0.01–0.1			Analyst **76**, 167 (1951)
	p-nitrosodimethylaniline	525	0.2–1.0	pH 1.2	Pt^{IV}, Au^{III}, Ag^+, CN^-, I^-	Anal. Chem. **26**, 1335 (1954)
	phenyl-α-pyridylketoxime	340	1.5–8		Au^{III}, CN^-	Vestn. Moscow Univ. **5**, 83 (1954)
	tetramethyldiaminodiphenyl-methane	570–600	>0.2	PbO_2 after electrolytical deposition	Cl^-, PO_4^{3-}	Z. Anal. Chem. **113**, 161 (1938)
Pd^{2+}	dimethylglyoxime			pH 1–2; extr. $CHCl_3$		Zh. Anal. Khim. **24**, 261 (1969)
	2,2-dipyridylglyoxime					Talanta **16**, 1330 (1969)
						Anal. Letters **2**, 167 (1969)
	8-hydroxyquinoline	435		$CHCl_3$		Bunseki Kagaku **18**, 136 (1969)
Pt^{IV}	diphenylthiosemicarbazide	750		n-butanol		Khim. Technol. **9**, 76 (1967)
Re^{VII}	2,4-diphenylthiosemi-carbazide	510	0.2–2.0	extr. $CHCl_3$		Z. Anal. Chem. **151**, 401 (1956); **152**, 96 (1956)
	α-furildioxime	532	0.4–6	1.2 M KCl; extr. $CHCl_3$ $SnCl_2$	Cu^{2+}, Mo^{2+}, Pd^{2+}	Vestn. Moscow Univ. **15**, 59 (1966)
						Anal. Chem. **240**, 19 (1968)
	thiourea	400	>0.8	2.5 N HCl		Zh. Anal. Khim. **10**, 228 (1955)

† Note: Roman numbers in this column refer to the formal oxidation state of the atom.

TABLE 29 (continued)

Metal ion	Reagent	Absorption maximum, nm	Concentration range, μg/ml	Reaction conditions	Interfering ions†	Refs.
Ru^{IV}	dithiooxamide	650, 654	1–5		Ir^{II}, Os^{IV}, Rh^{IV}	Anal. Chem. **22**, 1281 (1950); Anal. Chim. Acta **40**, 538 (1968)
	1,10-phenanthroline	660–700, 448	0.08–2.0, 0.3–1.6		Os^{IV}, Fe^{2+}	Anal. Chem. **29**, 1412 (1957)
	thiourea	620	2–15	0.1–4 N HCl	Co^{3+}, Cr^{3+}, Os^{IV}, Pd^{2+}	Anal. Chem. **22**, 1277 (1950)
Rh^{IV}	PAN					Anal. Letters **1**, 267 (1968); Zavod. Lab. **34**, 388 (1968)
Sb^{III}	thiourea	365	0.2–12	3 N H_2SO_4		Z. Anal. Chem. **143**, 81 (1954)
Sb^V	Methyl Violet	530, 615	0.1–2.0	HCl (1 : 8); extr. toluene or benzene	Tl, Cu^{2+}	Anal. Chim. Acta **7**, 462 (1952); **11**, 82 (1954)
	Rhodamine B	565	0.1–0.6	2–5 N acid	Au^{III}, Cr^{VI}, Fe^{3+}, Ga^{3+}, Hg^{2+}, Tl^{3+}, W^{VI}	Anal. Chim. Acta **13**, 455 (1955); Anal. Chem. **40**, 1146 (1968)
Sc^{3+}	alizarinsulfonic acid sodium salt	520	0.3–4			Anal. Chem. **27**, 1551 (1955)
	8-hydroxyquinoline	378–400	1–4	pH 10–10.4		Anal. Chim. Acta **16**, 334 (1957)

Ion	Reagent	λ	Range	pH	Ions	Reference
Sn^{2+}	Dithiol	530	2–16	pH < 4.0	Bi^{3+}, Cu^{2+}, Ni^{2+}, Co^{2+}, Ag^+, Hg^{2+}, Cd^{2+}, As^{3+}, F^-, PO_4^{3-}, NO_2^-	Anal. Chem. **26**, 735 (1954)
	2,3,7-trihydroxy-9-*p*-nitrophenyl-6-fluorone (Nitrophenyl-fluorone)	510	0.4–1.6	pH 1.8	Fe^{3+}, Ga^{3+}, Ge^{IV}, Ta^V, Zr^{VI}	Anal. Chem. **28**, 1276 (1956)
Sr^{2+}	Murexide	525	0–10			Trudi IREA **21**, 48 (1956)
Ta^V	phenylfluorone	530	0.4–2.5	pH ~ 7	Pt^{IV}, U^{VI}, Mo^{VI}, W^{VI}, Sb^V	Anal. Chem. **31**, 904 (1959)
	pyrogallol	325	40	4 N HCl		Anal. Chem. **25**, 1803 (1953); **30**, 465 (1958); Analyst **79**, 345 (1955); **80**, 194 (1955)
	pyrogallol		40.0	pH 0.5–2.0		Z. Anal. Chem. **241**, 186 (1968)
Th^{IV}	Arsenazo III	665	0.06–0.7	4–10 N HCl	Zr^{IV}, U^{IV}	Zav. Lab. **26**, 412 (1960)
	morin	410	0.1–1	pH 2		Anal. Chem. **28**, 1402 (1956)
	quercetin	422	0.6–6	pH 2.7–3.5	Al^{3+}, Cr^{3+}, Fe^{3+}, Ni^{2+}, U^{VI}	Anal. Chem. **29**, 1426 (1957)
	Thoron	545	2–15	pH 0.3–1	U^{VI}, Zr^{IV}, F^-, PO_4^{3-}	Anal. Chem. **25**, 416 (1953); **29**, 963 (1957)

† Note: Roman numbers in this column refer to the formal oxidation state of the atom.

TABLE 29 (continued)

Metal ion	Reagent	Absorption maximum, nm	Concentration range, µg/ml	Reaction conditions	Interfering ions‡	Refs.
Ti^{IV}	ascorbic acid	355–360	0.1–25	pH 3.5–6	V^V, Mo^{VI}, PO$_4^{3-}$	Anal. Chem. **24**, 947 (1952)
	chromotropic acid	410	0.03–0.3	pH 2.5–3.0	Cr^{3+}, Fe^{3+}, V^V, Co^{2+}, Cu^{2+}, Ni^{2+}, F$^-$	Anal. Chim. Acta **6**, 7 (1952) Z. Anal. Chem. **163**, 412 (1958)
	hydrocinnamic acid	390	0.01–2.8	pH 2.5–6; extr. amyl alcohol	Fe, V, PO$_4^{3-}$, CrO$_4^{2-}$, WO$_4^{2-}$	Sci. a. Cult. **20**, 146 (1954)
	sulfosalicylic acid	445	0.2–20	pH 3–5		Z. Anal. Chem. **139**, 92 (1953) **143**, 112 (1954)
	Tiron	410	0.3–3	pH 4.3–9.6	Cr^{3+}, Cu^{2+}, Mo^{VI}, V^V, W^{VI}	Anal. Chem. **19**, 100 (1947) **20**, 1208 (1948)
Tl^I	Thionalid		>0.5	pH 7.0		Mikrochem. Festschr. H. Molisch 42 (1936)
Tl^{III}	Brilliant Green	630	0.05–4.0	0.08–0.16 N HCl	Sb^{3+}, Hg^{2+}, Au^{III}, Sn^{2+}, W^{VI}, Cr^{VI}, ClO$_4^-$	Zh. Anal. Khim. **11**, 585 (1956)
	Methyl Violet	530	>0.2	0.2–0.3 N acid extr. toluene	Sb^{3+}, Hg^{2+}, Au^{III}	Zh. Anal. Khim. **11**, 585 (1956) Rev. Roum. Chim. **12**, 969 (1967)

SUMMARIZING TABLES

Tl^{III}	Rhodamine B	565	0.03–1.3	2 N HCl; extr. benzene	Au^{III}, Fe^{3+}	Anal. Chem. **31**, 1680 (1959) Anal. Chim. Acta **9**, 393 (1953)
U^{VI}	acetylacetone	360–365	<70	pH 6.0–7.0; extr. butyl acetate	Ti^{4+}, Cr^{3+}, Fe^{3+}, Cu^{2+}, Mn^{2+}	Anal. Chim. Acta **28**, 227 (1963)
	Arsenazo III	660	1–25	pH 4.5–5.0	Th^{4+}, Zr^{IV}	Dokl. Akad. Nauk SSSR **127**, 1231 (1959); Radiokhim. **2**, 682 (1960); Bunseki Kagaku **18**, 208 (1969)
	dibenzoylmethane	405–416	2–9	pH 7.0	Al^{3+}, Cd^{2+}, Ca^{2+}, Co^{2+}, Cr^{3+}, Cu^{2+}, Fe^{3+}, Mg^{2+}	Anal. Chem. **25**, 1200 (1953); **26**, 1635 (1963)
		360	<5.0	pH ≈ 7	Mn^{2+}, Ni^{2+}, W^{VI}, Pb^{2+}, Zr^{IV}, Th^{4+}, Sn^{IV}, Ti^{4+}, V^{V}, Zn^{2+}	Anal. Chim. Acta **28**, 227 (1963)
	8-hydroxyquinoline	425–500	<80	pH 4.5–9.0; extr. CHCl$_3$	Zr^{IV}, Th^{4+}, Tl^{4+}, V^{V}	Zh. Anal. Khim. **17**, 554 (1962)
	mercaptoacetic acid	380	4–64	pH 8–11	Al^{3+}, Ti^{4+}, Th^{4+}, Zr^{IV}	Anal. Chem. **21**, 1093 (1949)
	1-(2-pyridylazo)-2-naphthol (PAN)	560	0.1–110	pH 5–10; extr. CHCl$_3$		Z. Anal. Chem. **171**, 256 (1959) Anal. Chim. Acta **22**, 479 (1960)

† Note: Roman numbers in this column refer to the formal oxidation state of the atom.

TABLE 29 (continued)

Metal ion	Reagent	Absorption maximum, nm	Concentration range, μg/ml	Reaction conditions	Interfering ions‡	Refs.
U^{VI} (cont.)	sodium diethyldithio-carbamate	380	0.4–20	pH 6.6–7.0		Sandell: Colometric determination of traces of metals
		496	0.4–20	pH 7.0	Fe^{3+}, Cu^{2+}, Mn^{2+}, Ti^{4+}, V^V	Z. Anal. Chem. **144**, 165 (1955)
V^{IV}	pyrocatechol	600	0.4–8	pH ≈ 5.0	Fe^{2+}, Ti^{4+}, Mn^{2+}, U^{VI}, W^{VI}	Microchem. **23**, 194 (1937)
V^V	aniline	390	>0.05	extr. ethyl acetate	Mo^{VI}	Rev. Minerva **1**, 16 (1931) Chem. Abstr. **28**, 7193 (1934)
	8-hydroxyquinoline	550	1–6	pH 3.5–4.5		Anal. Chem. **25**, 604 (1953)
		365	0–25	pH 2.4; extr. $CHCl_3$	Fe^{3+}	Microchim. Acta 701 (1954)
W^{VI}	dithiol + $SnCl_2$	600	1–4		Mo^{VI}	Analyst **76**, 710 (1951) Z. Anal. Chem. **161**, 66 (1958)
	8-hydroxyquinoline	365	0–100	pH 2.5–3.5; extr. $CHCl_3$		Anal. Chim. Acta **28**, 132 (1963)
	hydroquinone	478	1–25	conc. H_2SO_4	Cr^{3+}, Fe^{3+}, Mo^{VI}, Ru^{IV}, Ti^{4+}, V^V	Z. Anal. Chem. **114**, 170 (1938); Chem. Abstr. **35**, 1726 (1941)

	Reagent	λ (nm)	Range	Conditions	Ions	Reference
Zn^{2+}	Dithizone	520–540; 620	0.1–1.0	pH 6.0–10.0	Sn^{2+}, Bi^{3+}, Pt^{2+}, Pd^{2+}, Au^{III}, Ag^+, Hg^{2+}, Ir^{IV}, Pb^{2+}, Cd^{2+}	Z. Anal. Chem. **97**, 385 (1934); Z. Anal. Chem. **239**, 158 (1968)
	8-hydroxyquinoline	395	0–60	pH 9.4; extr. methyl isobutyl ketone		Sci. Rep. Inst. Tokuku Univ. **A12**, 334 (1960)
	Zincon	620	0.1–2.4	pH 8.5–9.5	Al^{3+}, Be^{2+}, Bi^{3+}, Cd^{2+}, Co^{2+}, Cu^{2+}, Fe^{3+}, Mn^{2+}, Mo^{VI}, Ni^{2+}, Ti^{4+}	Anal. Chem. **26**, 1345 (1954)
Zr^{IV}	alizarinsulfonic acid sodium salt	530–560	0.1–4.0	2 N HCl	many elements	Anal. Chem. **20**, 370 (1948); Analyst **93**, 802 (1968)
	Arsenazo I	550–572	0.2–5.0	pH 1.5–1.8	Fe^{3+}, Hf^{IV}, Mo^{VI}, Sn^{IV}, Th^{4+}, Ti^{4+}, U^{VI}	Zav. Lab. **27**, 795 (1961); **26**, 415 (1960); **27**, 798 (1961)
	Arsenazo III	650	0.1–6.0	2 N HCl	U^{VI}, Th^{4+}, Cr^{3+}, PO_4^{3-}, F^-	
	chloranilic acid	330–350		2 M $HClO_4$	Fe^{3+}, Hf^{IV}, U^{IV}, Th^{IV}, Mo^{VI}, Sn^{IV}, Ti^{VI}	Anal. Chem. **29**, 558 (1957)
		525	2×10^{-6} to 5×10^{-5} M	2 M $HClO_4$	Hf^{IV}, U^V, Th^{4+}, Sn^{IV}	

† Note: Roman numbers in this column refer to the formal oxidation state of the atom.

TABLE 29 (continued)

Metal ion	Reagent	Absorption maximum, nm	Concentration range, μg/ml	Reaction conditions	Interfering ions†	Refs.
Zr^{IV} (cont.)	morin	436	0–5	0.4–0.7 N HCl	Fe^{3+}, Cr^{3+}, Ni^{2+}, Co^{2+}	Anal. Chim. Acta **16**, 346 (1957)
	p-dimethylaminoazophenyl-arsonic acid	410–430	>0.1	pH 8.0	W^{VI}, Mo^{VI}, Ti^{4+}	Zh. Priklad. Khim. **10**, 1969 (1937)
	phenylfluorone	535	>0.3	0.4–0.5 N HCl	Ta^V, Nb^V, Sn^{IV}, Sb^{3+}, Ti^{4+}, F^-, PO_4^{3-}	Z. Anal. Chem. **159**, 63 (1957)
	Pyrocathechol Violet	625	0.1–1.4	pH 5.6; EDTA	Ta^V, Nb^V, Th^{4+}, U^{VI}, F^-, PO_4^{3-}	Meeting Amer. Chem. Soc. **19B**, 50 (1957)
	quercetin	440	0.2–2	0.5 M HCl	Cr^{3+}, Fe^{3+}, Ga^{3+}, Hf^{IV}, Mo^{VI}, Nb^V, Sc^{3+}, Sb^{3+}, Sn^{2+}, Ta^V, Th^{IV}, Ti^{4+}, W^{VI}, F^-, PO_4^{3-}	Z. Anal. Chem. **230**, 271 (1967) Anal. Chem. **25**, 1886 (1953)

† Note: Roman numbers in this column refer to the formal oxidation state of the atom.

Abbreviations: *ppt.*: precipitate; *extr.*: solvent extraction method; *I.E.C.*: Ind. Eng. Chem.; *J.N.B.S.*: Journal of National Bureau of Standards; *dissol.*: dissolution.

THERMAL STABILITIES AND GRAVIMETRIC FACTORS OF SOME OF THE MORE IMPORTANT METAL COMPLEXES UTILIZED IN GRAVIMETRIC ANALYSIS

One of the most important factors of the application of organic reagents in gravimetric analysis is the thermal stability of their metal complexes, i.e. a metal complex containing a considerable organic moiety must be capable of being dried at elevated temperature without decomposition. The thorough examinations of Erdey et al.[148] and of Duval and his thermoanalytical school[138] have resulted in series of derivatograms and thermogravimetric curves, which clearly reveal the temperature interval within which a given precipitate can be made constant in weight. Table 30 has been compiled on the basis of these data.

The Table contains the compositions (column 3) of the individual complexes in the temperature intervals (column 4) within which the weight constancy of the precipitates can be achieved most advantageously. The last column gives the gravimetric factors, i.e. the multiplication factor with which the metal content of a complex can be calculated from the weight of precipitate.

It should be noted at this point that the thermal stabilities of metal complexes containing organic ligands depend upon the conditions of precipitation, though to a smaller extent than in the case of inorganic compounds. The data given in the Table refer to complexes precipitated by the usual, well-proved methods which have been detailed in general in the analytical procedures given in this book.

TABLE 30. THERMAL STABILITIES AND GRAVIMETRIC FACTORS OF SOME IMPORTANT METAL COMPLEXES USED IN GRAVIMETRIC ANALYSIS

Central atom	Organic reagent	Composition of precipitate	Temperature range, °C	Gravimetric factor
Al(III)	8-hydroxyquinoline	$Al(C_9H_6NO)_3$	102–220	0.0587
Au(I)	thiophenol	$Au(C_6H_5S)$	<157	0.6434
Ca(II)	oxalic acid	$CaC_2H_4 \cdot H_2O$	<105	0.2743
	oxalic acid	CaC_2O_4	210–320	0.3129
Cd(II)	8-hydroxyquinoline	$Cd(C_9H_6NO)_2$	240–380	0.2805
	quinaldic acid	$Cd(C_{10}H_6NO_2)_2$	125–260	0.2461
Co(II)	8-hydroxyquinoline	$Co(C_9H_6NO)_2$	180–290	0.1697
	pyridine + SCN$^-$	$Co(C_5H_5N)_4(SCN)_2$	<70	0.1199
Co(III)	1-nitroso-2-naphthol	$Co(C_{10}H_6NO_2)_3$	130–200	0.1024
Cr(III)	8-hydroxyquinoline	$Cr(C_9H_6NO)_3$	70–150	0.1074
Cu(II)	anthranilic acid	$Cu(NH_2 \cdot C_6H_4CO_2)_2$	<225	0.1892
	8-hydroxyquinoline	$Cu(C_9H_6NO)_2$	66–269	0.1806
	oxalic acid	CuC_2O_4	100–270	0.4192
	quinaldic acid	$Cu(C_{10}H_6NO_2)_2 \cdot H_2O$	<120	0.1492
	salicylaldoxime	$Cu(C_7H_6NO_2)_2$	<150	0.1892
	Thionalid	$Cu(C_{12}H_{10}ONS)_2$	148–167	0.1281

TABLE 30 (continued)

Central atom	Organic reagent	Composition of precipitate	Temperature range, °C	Gravimetric factor
Fe(III)	Cupferron	Fe($C_6H_5N \cdot NO \cdot O$)$_3$	< 98	0.1195
	8-hydroxyquinoline	Fe(C_9H_6NO)$_3$	<284	0.1144
Ga(III)	5,7-dibromo-8-hydroxyquinoline	Ga($C_9H_4NOBr_2$)$_3$	100–224	0.07146
Hf(IV)	mandelic acid	Hf($C_6H_5CHOH \cdot CO_2$)$_4$	90–260	0.2279
Hg(II)	Thionalid	Hg($C_{12}H_{10}ONS$)$_2$	90–169	0.3168
In(III)	8-hydroxyquinoline	In(C_9H_6NO)$_3$	100–285	0.2098
K(I)	dipicrylamine	K($C_{12}H_4O_{12}N_7$)	50–220	0.08192
	sodium tetraphenylboron	K[B(C_6H_5)$_4$]	<120	0.1091
Mg(II)	8-hydroxyquinoline	Mg(C_9H_6NO)$_2$	155–210	0.07779
Mn(II)	8-hydroxyquinoline	Mn(C_9H_6NO)$_2$	117–250	0.1600
Mo(VI)	8-hydroxyquinoline	MoO$_2$(C_9H_6NO)$_2$	40–220	0.2305
Na(I)	UO$_2$(CH$_3$COO)$_2$ + Zn(CH$_3$COO)$_2$	NaZn(UO$_2$)$_3$ · (CH$_3$COO)$_9$ · 6.5 H$_2$O	105–120	0.01486
Ni(II)	dimethylglyoxime	Ni($C_8H_{14}O_4N_4$)	80–200	0.2032
	8-hydroxyquinoline	Ni(C_9H_6NO)$_2$	200–300	0.1692
	pyridine + SCN$^-$	Ni(C_5H_5N)$_4$(SCN)$_2$	< 60	0.1195
Pb(II)	salicylaldoxime	Pb($C_7H_5NO_2$)$_2$	45–180	0.4340
	Thionalid	Pb($C_{12}H_{10}ONS$)$_2$	71–134	0.3409
Pd(II)	dimethylglyoxime	Pd($C_4H_7O_2N_2$)$_2$	45–171	0.3161
	o-phenanthroline	PdCl$_2$ · $C_{12}H_8N_2$	50–389	0.2976
Rb(I)	sodium tetraphenylboron	Rb[B(C_6H_5)$_4$]	<240	0.2112
Re(VII)	Nitron	($C_{20}H_{16}NH$)ReO$_4$	91–288	0.3304
Sc(III)	8-hydroxyquinoline	Sc(C_9H_6NO)$_3$ · (C_9H_6NOH)	<125	0.0722
Sr(II)	oxalic acid	SrC$_2$O$_4$ · H$_2$O	< 60	0.4525
	oxalic acid	SrC$_2$O$_4$	200–270	0.4989
Th(IV)	8-hydroxyquinoline	Th(C_9H_6NO)$_4$ · (C_9H_6NOH)	< 80	0.2433
Ti(VI)	5,7-dichloro-8-hydroxyquinoline	TiO($C_9H_4NOCl_2$)$_2$	105–195	0.0978
Tl(III)	(C_6H_5)$_4$AsCl	(C_6H_5)$_4$AsTlCl$_4$	50–218	0.2802
U(VI)	8-hydroxyquinoline	UO$_2$(C_9H_6NO)$_2$ · (C_9H_6NOH)	<160	0.3384
W(VI)	8-hydroxyquinoline	WO$_2$(C_9H_6NO)$_2$	<140	0.3647
Zn(II)	anthranilic acid	Zn($C_7H_6NO_2$)$_2$	<240	0.1936
	8-hydroxyquinoline	Zn(C_6H_6NO)$_2$	180–250	0.18485
	quinaldic acid	Zn($C_{10}H_6NO_2$)$_2$ · H$_2$O	<150	0.15285
Zr(IV)	mandelic acid	Zr($C_8H_7O_3$)$_4$	60–180	0.1311
	p-bromomandelic acid	Zr(BrC$_8H_6O_3$)$_4$	<150	0.0902

STABILITY CONSTANTS OF THE CHELATES OF COMPLEXING AGENTS USED AS ANALYTICAL REAGENTS

It is clear from the subject matter of this book that a knowledge of the stability constants of the metal complexes in a system may be of tremendous help in the elaboration of analytical procedures based on complex formation. Without a knowledge of these stability data, analytical procedures may be established merely by empirical trial and error. In contrast, a knowledge of the stability data makes possible the scientific design of the analytical procedure, selection of the optimum reaction conditions, pH, masking agent and its concentration, etc. In addition, the stability constants of the complexes formed by the ligand used as analytical reagent with various metal ions also reveal in many cases those metal ions which interfere with the determination of a given metal ion.

Nevertheless, even today there are practical analysts who prefer the *empirical* method in the elaboration of analytical procedures. One of the reasons for this is that many of the stability constants reported in the literature are unreliable. Cases are known where data given by different authors for the stability of the same complex differ by several orders of magnitude.[433] The analyst cannot usually undertake revision of such divergent results, and he rather disregards them. It is a complex chemical task to review critically the stability data published in the literature and to compile critical stability tables from which the analyst can gain the necessary data.

In general, the errors in stability constants originate from various sources.[37] For instance:

1. Improper choice of the measuring method.
2. The evaluation is inconsistent with the experimental data.
3. Application of incorrect auxiliary constants, e.g. acidic dissociation constants.
4. Inaccurate experimental work.

The first three sources of error can be recognized in many cases on reading the relevant publication. In this way a certain selection of the data available in the literature can be achieved without reproducing the experimental work. It is the intention of the following considerations to provide assistance in this.

1. In the measurement of equilibrium constants we always measure the changes in concentration (activity) of species (metal ion, complex or free ligand) in equilibrium. The measurement of concentration may be accomplished by essentially any analytical procedure which does not alter the equilibrium in question during the procedure. Such methods are almost exclusively instrumental. The inexperienced analyst may make the error of utilizing data provided by the instrument, even if the data do not correspond to changes in concentration of the species concerned (metal ion or complex).

For instance, a certain potential value will of course be measurable by means of an electrode dipped into a solution containing metal ions. However, the electrode may be used for the measurement of the metal ion concentration only if the electrode reaction is reversible. Thus, for example, a nickel electrode may never be used for the measurement of equilibrium data of nickel(II) complexes, because the $Ni^{2+} + 2e^- \rightarrow Ni^0$ reaction is irreversible on the nickel electrode. Therefore the concentration of nickel ions calculated from the electrode potential of such a system will necessarily differ from the actual concentration of nickel ions in the solution.

In polarographic measurements on equilibria too it is an important requirement that the measured wave be a reversible diffusion wave.

Extinction measurement data may be utilized to follow complex equilibria only with the knowledge of which complex or complexes absorb at the wavelength of measurement. However, the determination of these data, particularly in case of many, successively-formed complexes, is a rather difficult task. Examinations based on extinction measurements are intrinsically complicated because of the need to know the extinction coefficients of all the complex species absorbing at the wavelength of measurement.

The following must be considered as a general principle: whenever the actual relationship between the experimental data and the concentrations of the complexes determining them is unknown (the calculation is made on the basis of some presumed relationship) there is no reason to expect current results. Accordingly, even the most thorough experimental work fails to provide reliable stability constants if the experimental method is unsuitable.

2. The proper choice of the evaluation method is determined by the system examined and the nature of the experimental results.

In general, the experimental data reflect all equilibria existing in the given system, not only those whose constants are to be determined. In an exact, e.g. computer-performed, evaluation of the experimental data, all equilibria can be taken into account. However, in general this is not necessary. Consideration of all the possible equilibria would make the calculation unnecessarily complicated without greatly increasing the reliability of the constants obtained. Certain simplifications, such as the disregarding of equilibria which are negligible under the given conditions and within the concentration range in question, are therefore necessary. The error of evaluation usually originates from inadmissible neglections and over-simplifications.

For instance, a frequently occurring error is the disregard of acidic dissociation equilibria of chelating ligands in acidic solutions, which are negligible in neutral or alkaline media only. This had been the case, for instance, with complexes of EDTA, where only the acidic dissociation constants of the four carboxylic groups used to be taken into account, until Beck[38] pointed out that in acidic media the two nitrogen atoms also undergo protonation.

It also occurs frequently that the buffer applied for adjustment of the pH of the solution to the desired value contains a ligand (ammonia, acetate, etc.) which forms a complex with the metal ion to be determined. A disregard of the equilibria involving these complexes may cause an error of several orders of magnitude in the equilibrium constants of the system examined.

Another error which occurs rarely today, but which was frequent in the equilibrium data measured in the fifties and earlier, is the application of Bjerrum's "half-value method"[59] in the evaluation of the experimental data of systems in which the relation between successive stability constants excludes the utilization of this method. Bjerrum himself called attention during elaboration of the procedure to the requirement that the method provides correct results only in the case of the relation $K_n : K_{n+1} \gg 1$ between the successive stability constants. In dimethylglyoxime complexes strong intramolecular hydrogen bonds stabilize the MA_2 complex, and hence $K_2 > K_1$.[89, 142] Nevertheless, many authors have used the "half-value method" in the evaluation of the experimental data.[65, 112, 465] Of course, under such circumstances even the most thorough measurements can provide results which are at most only of approximate accuracy, and are useful merely as informatory data.

3. In examinations of complex equilibria we almost always need a knowledge of certain "auxiliary constants", such as the acidic dissociation constants of the ligand, the stability constants of complexes of accompanying ligands, extinction coefficients, distribution coefficients, etc. An error in such an auxiliary constant may lead to a result incorrect by several orders of magnitude, despite the fact that the measurement was made with the most suitable method, with maximum accuracy, and the data were appropriately evaluated. Therefore, in deciding the reliability of a given stability constant attention should also be devoted to the soundness of the auxiliary constants involved.

4. Finally a few words about the reliability of the experimental work, which is the basic condition of the applicability of any data. Apart from reproducing the measurements, this is most difficult to check. What can be seen from the literary source is whether the experiments were carried out under well-defined conditions (constant temperature, ionic strength, solvent composition, concentration, etc.). Some information can also be gained from the standard deviation of the measurements and the quality of the reagents applied.

With the above intentions in mind, the author has surveyed the literature concerning the measurements on the equilibria of the complexes of 28 ligands used as analytical reagents. Efforts were made in each case to select the most probable value from among the many contradictory ones published. Tables 31–58 have been compiled on the basis of these considerations. In general, only one stability constant has been given for each metal complex. Exceptions were made only in those cases where it appeared reasonable to include two data measured under different conditions (e.g. in different solvents). The Tables include the protonation constants (reciprocal acidic dissociation constants) involved in the calculations. In some cases, when the stability constants given in the Tables were calculated with different protonation constants, each of these latter is shown. In these cases it is quite clear from the references as well as the specification of the medium and the ionic strength which values relate to one another.

In addition to the stability constants, the measurement method and the experimental conditions (medium, ionic strength, temperature) are also indicated. In order to facilitate the gaining of further information, references have been made to the original publications.

The author sincerely hopes that these Tables will efficiently aid analytical chemists in the solution of their metal analytical problems.

SIGNS AND ABBREVIATIONS USED

1. Equilibrium constants:

$$K_1 = \frac{[MA]}{[M][A]}; \quad K_2 = \frac{[MA_2]}{[MA][A]}; \quad K_n = \frac{[MA_n]}{[MA_{n-1}][A]};$$

$$\beta_n = K_1 K_2 \ldots K_n = \frac{[MA_n]}{[M][A]^n}.$$

2. Method of measurement:
   ext.: solvent extraction
   photometr.: spectrophotometry
   calorim.: calorimetry
   sol.: solubility
   pH-metric (or pH-metr.): pH-metry, carried out with the glass electrode
   pot.: some other potentiometric method (usually with indication of the electrode)
   polar.: polarography
3. Medium, ionic strength: Ionic strength and the salt used for the adjustment of the ionic strength, e.g. 0.1 KNO_3.
   → 0: extrapolated to zero ionic strength.

TABLE 31. STABILITY CONSTANTS OF ACETYLACETONE COMPLEXES

Central atom	$\log K_1$	$\log K_2$	$\log K_3$	$\log K_4$	Measuring method	Ionic strength of medium	Temp., °C	Refs.
H^+	8.95				pH-metric	$\to 0$	30	1, 2
Al^{3+}	8.6	7.9	5.8		pH-metric	$\to 0$	30	2
Be^{2+}	7.8	6.7			pH-metric	$\to 0$	30	3
Cd^{2+}	3.83	2.76			pH-metric	$\to 0$	30	1, 3, 4
Ce^{3+}	5.28	3.98	3.38		pH-metric	$\to 0$	30	1, 4
Co^{2+}	5.4	4.11			pH-metric	$\to 0$	30	1, 3, 4
Cu^{2+}	8.22	6.73			pH-metric	$\to 0$	30	1, 3
Dy^{3+}	6.03	4.67	3.34		pH-metric	0.1	30	5
Er^{3+}	5.99	4.68	3.38		pH-metric	0.1	30	5
Eu^{3+}	5.87	4.48	3.29		pH-metric	0.1	30	5
Fe^{2+}	5.07	3.60			pH-metric	$\to 0$	30	1, 3
Fe^{3+}	9.8	9.0	7.4		pH-metric	$\to 0$	30	2
Ga^{3+}	9.5	8.4	5.7		pH-metric	$\to 0$	30	2
Gd^{3+}	5.9	4.48	3.41		pH-metric	0.1	30	5
Ho^{3+}	6.05	4.68	3.40		pH-metric	0.1	30	5
In^{3+}	8.0	7.1			pH-metric	$\to 0$	30	2
La^{3+}	4.96	3.45	2.5		pH-metric	0.1	30	5
Lu^{3+}	6.23	4.77	3.63		pH-metric	0.1	30	5
Mg^{2+}	3.63	2.54			pH-metric	$\to 0$	30	1, 3, 4
Mn^{2+}	4.18	3.07			pH-metric	$\to 0$	30	1, 3, 4
Nd^{3+}	5.30	4.1	3.2		pH-metric	0.1	30	5
Ni^{2+}	5.92	4.46	2.11		pH-metric	$\to 0$	30	2
Pd^{2+}	16.2	10.9			pH-metric	$\to 0$	30	6
Pr^{3+}	5.4	4.1	3.0		pH-metric	$\to 0$	30	2
Sc^{3+}	8.0	7.2			pH-metric	$\to 0$	30	2
Sm^{3+}	5.59	4.46	2.9		pH-metric	0.1	30	5
Tb^{3+}	6.02	4.61	3.41		pH-metric	0.1	30	5
Th^{4+}	8.8	7.4	6.3	4.2	pH-metric	$\to 0$	30	2
Tm^{4+}	6.09	4.76	3.48		pH-metric	0.1	30	5
U^{6+}	9.02	8.26	6.52	5.98	extr.	0.1 $NaClO_4$/benzene	30	7
UO_2^{2+}	7.74	6.43			pH-metric	$\to 0$	30	1, 3
Y^{3+}	5.87	4.98	3.25		pH-metric	0.1	30	5
Yb^{3+}	6.18	4.86	3.60		pH-metric	0.1	30	5
Zn^{2+}	4.98	3.83			pH-metric	$\to 0$	30	1, 3

1. IZATT, R. M., HAAS, C. G., BLOCK, B. P. and FERNELIUS, W. C.: *J. Phys. Chem.* **58,** 1133 (1954).
2. IZATT, R. M., FERNELIUS, W. C., HAAS, C. G. and BLOCK, B. P.: *J. Phys. Chem.* **59,** 170 (1955).
3. IZATT, R. M., FERNELIUS, W. C. and BLOCK, B. P.: *J. Phys. Chem.* **59,** 80 (1955).
4. IZATT, R. M., FERNELIUS, W. C. and BLOCK, B. P.: *J. Phys. Chem.* **59,** 235 (1955).
5. GRENTHE, J., FERNELIUS, W. C.: *J. Amer. Chem. Soc.* **82,** 6258 (1960).
6. DROLL, H. A., BLOCK, B. P. and FERNELIUS, W. C.: *J. Phys. Chem.* **61,** 100 (1957).
7. RYDBERG, J.: *Acta Chem. Scand.* **14,** 157 (1960).

TABLE 32. STABILITY CONSTANTS OF α,α'-DIPYRIDYL COMPLEXES

Central atom	$\log K_1$	$\log K_2$	$\log K_3$	Measuring method	Ionic strength of medium	Temperature, °C	Ref.
H^+	4.49			pH-metric	0.1 $NaNO_3$	20	1
Cd^{2+}	4.25	3.6	2.7	pH-metric	0.1 $NaNO_3$	20	1
Co^{2+}	6.06	5.36	4.60	pH-metric	0.1 $NaNO_3$	20	1
Cu^{2+}	8.0	5.60	3.48	pot. Hg + glass-electr.	0.1 $NaNO_3$	20	1
Fe^{2+}	$K_3 > K_1 K_2$	$\log \beta_3$ 17.45		pH-metric	0.1 $NaNO_3$	20	1
Hg^{2+}	9.64	7.10	2.8	pot. Hg + glass-electr.	0.1 $NaNO_3$	20	1
Ni^{2+}	7.13	6.88	6.53	pot. Hg + glass-electr.	0.1 $NaNO_3$	20	1
Pb^{2+}	2.9			pH-metric	0.1 $NaNO_3$	20	1
Zn^{2+}	5.30	4.53	3.80	pot. Hg + glass-electr.	0.1 $NaNO_3$	20	1

1. ANDEREGG, G.: *Helv. Chim. Acta* **46**, 2397 (1963).

TABLE 33. STABILITY CONSTANTS OF α-FURILDIOXIME COMPLEXES

Central atom	$\log K_1$	$\log K_2$	Measuring method	Ionic strength of medium	Temperature, °C	Ref.
H^+	11.1	11.4	pH-metric	75% dioxane, 0.1 $NaClO_4$	25	1
Co^{2+}	8.2	7.2	pH-metric	75% dioxane, 0.1 $NaClO_4$	25	1
Cu^{2+}	$K_1 \ll K_2$	$\log \beta_2$ 18.6	pH-metric	75% dioxane, 0.1 $NaClO_4$	25	1
Fe^{2+}	$K_1 \ll K_2$	$\log \beta_2$ 21.8	pH-metric	75% dioxane, 0.1 $NaClO_4$	25	1
Mn^{2+}	6.4	5.3	pH-metric	75% dioxane, 0.1 $NaClO_4$	25	1
Ni^{2+}	6.9	7.2	pH-metric	75% dioxane, 0.1 $NaClO_4$	25	1
Zn^{2+}	$K_1 \ll K_2$	$\log \beta_2$ 15.7	pH-metric	75% dioxane, 0.1 $NaClO_4$	25	1

1. BURGER, K. and PAPP-MOLNÁR, E.: *Acta Chim. Hung.* **53**, 111 (1967).

TABLE 34. STABILITY CONSTANTS OF α-NITROSO-β-NAPHTHOL COMPLEXES

Central atom	$\log K_1$	$\log K_2$	$\log K_3$	Measuring method	Ionic strength of medium	Temperature, °C	Refs.
H^+	11.62			pH-metric	75% dioxane	30	1
Ag^+	7.74			pH-metric	75% dioxane	30	1
Co^{2+}	10.67	12.14		pH-metric	75% dioxane	30	1
Cu^{2+}	12.52	10.85		pH-metric	75% dioxane	30	1
Mg^{2+}	6.05	4.72		pH-metric	75% dioxane	30	1
Nd^{3+}	9.5	8.2	7.86	pH-metric	75% dioxane	30	1
Ni^{2+}	10.75	10.54	6.80	pH-metric	75% dioxane	30	1
Pb^{2+}	9.73	7.58		pH-metric	75% dioxane	30	1
Pr^{3+}	9.04	8.02	6.79	pH-metric	75% dioxane	30	1
Th^{4+}	8.50	7.63		extr.	0.1 Na^+, H^+, ClO_4^-	25	2
Y^{3+}	9.02	8.72	7.30	pH-metric	75% dioxane	30	1
Zn^{2+}	9.32	7.70		pH-metric	75% dioxane	30	1

1. CALLAHAN, C. M., FERNELIUS, W. C. and BLOCK, B. P.: *Anal. Chim. Acta* **16**, 101 (1957).
2. DYRSSEN, D., DYRSSEN, M. and JOHANSSON, E.: *Acta Chem. Scand.* **10**, 106 (1956).

TABLE 35. STABILITY CONSTANTS OF CDTA† COMPLEXES

Central atom	$\log K_1$	Measuring method	Ionic strength of medium	Temperature, °C	Refs.
H^+	11.70	potentiometric, H electr.	0.1 KCl	20	1
	$\log K_2$ 6.12				
	$\log K_3$ 3.52				
	$\log K_4$ 2.43				
Al^{3+}	17.63	polarography	0.1 KNO_3	20	2
Ba^{2+}	8.64	pH-metric, calorimetric	0.1 KNO_3	20	3
Ca^{2+}	13.15	pH-metric, calorimetric	0.1 KNO_3	20	3
Cd^{2+}	19.88	pH-metric, calorimetric	0.1 KNO_3	20	3
Ce^{3+}	16.76	polarography	0.1 KNO_3	20	2
Co^{2+}	19.57	pH-metric, calorimetric	0.1 KNO_3	20	3
Cu^{2+}	21.95	pH-metric, calorimetric	0.1 KNO_3	20	3
Dy^{3+}	19.69	polarography	0.1 KNO_3	20	2
Er^{3+}	20.68	polarography	0.1 KNO_3	20	2
Eu^{2+}	18.77	potentiometric, Hg electr.	0.1 KNO_3	25	4
Eu^{3+}	18.62	polarography	0.1 KNO_3	20	2
Fe^{3+}	10.95	(FeH_2A) photometric		20	5
Ga^{3+}	22.91	polarography	0.1 KNO_3	20	2
Gd^{3+}	18.77	polarography	0.1 KNO_3	20	2
Hg^{2+}	24.95	pH-metric, calorimetric	0.1 KNO_3	20	3
La^{3+}	16.91	pH-metric, calorimetric	0.1 KNO_3	20	3
Lu^{3+}	21.51	polarography	0.1 KNO_3	20	2
Mg^{2+}	10.97	pH-metric, calorimetric	0.1 KNO_3	20	3
Mn^{2+}	17.43	pH-metric, calorimetric	0.1 KNO_3	20	3
Nd^{3+}	17.68	polarography	0.1 KNO_3	20	2
Ni^{2+}	19.4	potentiometric, Hg electr.	0.1 KNO_3	20	6
Pb^{2+}	20.33	pH-metric, calorimetric	0.1 KNO_3	20	3
Pr^{3+}	17.31	polarography	0.1 KNO_3	20	2
Sm^{3+}	18.38	polarography	0.1 KNO_3	20	2
Sr^{2+}	10.54	pH-metric, calorimetric	0.1 KNO_3	20	3
Tb^{3+}	19.50	polarography	0.1 KNO_3	20	2
Tm^{3+}	20.96	polarography	0.1 KNO_3	20	2
VO^{2+}	19.40	polarography	0.1 KNO_3	20	2
Y^{3+}	19.15	polarography	0.1 KNO_3	20	2
Yb^{3+}	21.12	polarography	0.1 KNO_3	20	2
Zn^{2+}	19.32	pH-metric, calorimetric	0.1 KNO_3	20	3

1. SCHWARZENBACH, G. AND ACKERMANN, H.: *Helv. Chim. Acta* **32**, 1682 (1949).
2. SCHWARZENBACH, G., GUT, R. and ANDEREGG, G.: *Helv. Chim. Acta* **37**, 937 (1954).
3. ANDEREGG, G.: *Helv. Chim. Acta* **46**, 1833 (1963).
4. MOELLER, T. and HSEN, T. M.: *J. Inorg. Nucl. Chem.* **24**, 1635 (1962).
5. BECK, M. T. and GÖRÖG, S.: *Acta Chim. Hung.* **22**, 159 (1960).
6. HOLLOWAY, J. H. and REILLY, C. N.: *Anal. Chem.* **32**, 249 (1960).

† CDTA = *trans*-cyclohexane-1,2-diamine-N,N,N',N'-tetraacetate.

TABLE 36. STABILITY CONSTANTS OF DIETHYLDITHIOCARBAMATE COMPLEXES

Central atom	$\log \beta_2$	$\log \beta_3$	Measuring method	Ionic strength of medium	Temperature, °C	Ref.
Co^{2+}		14.40	photometric	Water–methanol 1:3 mixture 0.3 M $NaClO_4$	25	1
Fe^{2+}		11.34	photometric	Water–methanol 1:3 mixture 0.3 M $NaClO_4$	25	1
Ni^{2+}	8.56		photometric	Water–methanol 1:3 mixture 0.3 M $NaClO_4$	25	1

1. PAPP-MOLNÁR, E., VÁSÁRHELYI-NAGY, H. and BURGER, K.: *Acta Chim. Hung.* **64**, 317 (1970).

TABLE 37. STABILITY CONSTANTS OF DIISOPROPYLDITHIOPHOSPHATE COMPLEXES

Central atom	$\log \beta_2$	Measuring method	Ionic strength of medium	Temperature, °C	Ref.
Cu^{2+}	8.68	photometric	Water–methanol 1:3 mixture 0.3 M $NaClO_4$	25	1
Ni^{2+}	5.24	photometric	Water–methanol 1:3 mixture 0.3 M $NaClO_4$	25	1

1. BURGER, K., PAPP-MOLNÁR, E., VÁSÁRHELYI-NAGY, H. and KORECZ, L.: *Acta Chim. Hung.* **64**, 323 (1970).

TABLE 38. STABILITY CONSTANTS OF DIMETHYLGLYOXIME COMPLEXES

Central atom	$\log K_1$	$\log K_2$	Measuring method	Ionic strength of medium	Temperature, °C	Refs.
H^+	12.83		pH-metric	50% dioxane, 0.3 $NaClO_4$	25	1
H^+	10.45		sol. measmut. pH-metric	0.1	25	2
Co^{2+}	10.80	10.2	pH-metric	50% dioxane, 0.3 $NaClO_4$	25	3
Co^{2+}	8.35	8.63	pH-metric	0.1	25	4
Cu^{2+}	11.0	12.5	pH-metric	50% dioxane, 0.3 $NaClO_4$	25	3
Cu^{2+}	$K_1 < K_2$	$\log \beta_2$ 19.24	extr.	0.1 $NaClO_4$, $CHCl_3$	25	5
Fe^{2+}	10.8	9.2	pH-metric	50% dioxane, 0.3 $NaClO_4$	25	3
Mn^{2+}	8.6	8.6	pH-metric	50% dioxane, 0.3 $NaClO_4$	25	3
Ni^{2+}	10.62	11.08	pH-metric	50% dioxane	25	2
Ni^{2+}	$K_1 < K_2$	$\log \beta_2$ 17.24	extr.	0.1 $NaClO_4$, $CHCl_3$	25	6
Pd^{2+}	$K_1 < K_2$	$\log \beta_2$ 34.1	extr.	0.1 $NaClO_4$, $CHCl_3$	25	7
Zn^{2+}	11.0	12.5	pH-metric	50% dioxane, 0.1 $NaClO_4$	25	3

1. BOCHKOVA, V. M. and PESHKOVA, V. M.: *Zh. Neorg. Khim.* **3**, 1131 (1958).
2. DYRSSEN, D.: *Trans. Roy. Inst. Technology, Stockholm* **26**, 220 (1964).
3. BURGER, K. and RUFF, I.: *Talanta* **10**, 329 (1963).
4. BANKS, C. V. and ANDERSON, S.: *Inorg. Chem.* **2**, 112 (1963).
5. DYRSSEN, D. and HENNICHS, M.: *Acta Chem. Scand.* **15**, 47 (1961).
6. DYRSSEN, D., KRASOVEC, F. and SILLEN, L. G.: *Acta Chem. Scand.* **13**, 50 (1959).
7. BURGER, K. and DYRSSEN, D.: *Acta Chem. Scand.* **17**, 1489 (1963).

TABLE 39. STABILITY CONSTANTS OF DITHIOOXAMIDE (RUBEANIC ACID) COMPLEXES

Central atom	$\log K_1$	$\log K_2$	$\log \beta_2$	Measuring method	Ionic strength of medium	Temperature, °C	Ref.
H^+	12.92				75% dioxane, 0.1 M $NaClO_4$	20	1
Mn^{2+}	6.29	7.02	13.31	pH-metric	75% dioxane, 0.1 M $NaClO_4$	20	1
Fe^{2+}	14.13	12.75	26.88	photometric	75% dioxane, 0.1 M $NaClO_4$	20	1
Co^{2+}	17.71	13.71	31.42	photometric	75% dioxane, 0.1 M $NaClO_4$	20	1
Ni^{2+}	$K_1 < K_2$		20.95 21.01	pH-metric photometric	75% dioxane, 0.1 M $NaClO_4$	20	1
Cu^{2+}	$K_1 < K_2$		30.22	photometric	75% dioxane, 0.1 M $NaClO_4$	20	1
Zn^{2+}	8.68	6.68	15.36	pH-metric	75% dioxane, 0.1 M $NaClO_4$	20	1

1. BURGER, K., SZÁNTÓ-HORVÁTH, G. and PAPP-MOLNÁR, E.: *Acta Chim. Hung.* **71**, 127 (1972)

Table 40. Stability Constants of EDTA Complexes

Central atom	$\log K_1$	Measuring method	Ionic strength of medium	Temperature °C	Refs.
H^+	10.26	pH-metric	0.1 KNO_3	20	1
	$\log K_2$ 6.16				
	$\log K_3$ 2.67				
	$\log K_4$ 1.99				
Ag^+	7.32	pH-metric	0.1 KNO_3	20	2
Al^{3+}	16.13	polarography	0.1 KNO_3	20	1
Ba^{2+}	7.76	pot. H electrode	0.1 KCl	20	
Ca^{2+}	10.85	pot. Hg electrode	0.1 $NaNO_3$	21.7	4
Cd^{2+}	16.61	pot. Hg electrode	0.1 $NaNO_3$	21.7	4
Co^{2+}	16.31	polarography	0.1 KNO_3	21.7	1
Co^{3+}	36	pot. redox electrode	0.1 KCl	20	6
Cu^{2+}	18.79	pH-metric	0.1 KCl	20	1
Dy^{3+}	18.00	pH-metric	0.1 KCl	20	1
Er^{3+}	18.98	polarography	0.1 KNO_3	20	1
Eu^{3+}	17.35	polar, pH-metric	0.1 KNO_3	20	1
Fe^{2+}	14.33	pot. H electrode	0.1 KCl	20	1, 5, 6
Fe^{3+}	24.23	photometric	$\rightarrow 0$	20	7
Ga^{3+}	20.27	polarography	0.1 KNO_3	20	1
Gd^{3+}	17.10	pH-metric	0.1 KCl	20	1
Hg^{2+}	21.80	pot. Hg electrode	0.1 KNO_3	20	1
Ho^{3+}	18.74	polarography	0.1 KNO_3	20	1
In^{3+}	24.95	polarography	0.1 KNO_3	20	2
La^{3+}	15.50	polarography	0.1 KNO_3	20	1
Li^+	2.79	pot. H electrode	0.1 KCl	20	3
Lu^{3+}	19.83	pH-metric	0.1 KNO_3	20	1
Mg^{2+}	8.69	pot. H electrode	0.1 KCl	20	1
Mn^{2+}	14.04	polarography	0.1 KNO_3	20	1
Mo^{5+}	6.36	photometric			8
Na^+	1.66	pot. H electrode	0.1 KCl	20	3
Ni^{2+}	18.56	pH-metric	0.1 KCl	20	1
Pb^{2+}	18.04	polarography	0.1 KNO_3	20	1
Pr^{3+}	16.40	polarography	0.1 KNO_3	20	1
Sc^{3+}	23.1	polarography	0.1 KNO_3	20	1
Sm^{3+}	16.9	pH-metric	0.1 KCl	20	1
Sr^{2+}	8.63	pot. H electrode	0.1 KCl	20	1
Th^{3+}	17.6	pH-metric	0.1 KCl	20	1

TABLE 40 *(continued)*

Central atom	log K_1	Measuring method	Ionic strength of medium	Temperature, °C	Refs.
Th^{4+}	23.2	polarography	0.1 KNO_3	20	1
Ti^{3+}	17.3	polarography	0.1 $NaNO_3$	25	9
Tl^{3+}	22.5	pH-metric	0.1	20	10
Tm^{3+}	19.32	polarography	0.1 KNO_3	20	1
V^{2+}	12.70	pH-metric	0.1 KCl	20	11
V^{3+}	25.9	pot. redox electrode	0.1 KCl	20	11
VO^{2+}	18.77	polarography, pot. redox electrode	0.1 KCl	20	11
V^{5+}	18.05	photometric	0.1 $NaClO_4$	25	12
Y^{3+}	18.08	polarography	0.1 KNO_3	20	1
Yb^{3+}	19.51	polarography	0.1 KNO_3	20	1
Zn^{2+}	16.50	polarography	0.1 KNO_3	20	1
Zr^{4+}	19.40	pH-metric	0.1 $NaClO_4$	25	13

1. SCHWARZENBACH, G., GUT, R. and ANDEREGG, G.: *Helv. Chim. Acta* **37**, 937 (1954).
2. Schwarzenbach's data; SILLEN, L. G. and MARTELL, A. E.: *Stability Constants*. The Chemical Society, London, 1964.
3. SCHWARZENBACH, G. and ACKERMANN, H.: *Helv. Chim. Acta* **30**, 1798 (1947).
4. SCHWARZENBACH, G. and ANDEREGG, G.: *Helv. Chim. Acta* **40**, 1773 (1957).
5. SCHWARZENBACH, G. AND HELLER, J.: *Helv. Chim. Acta* **34**, 576 (1951).
6. SCHWARZENBACH, G. AND FREITAG, E.: *Helv. Chim. Acta* **34**, 1503 (1951).
7. BECK, M. T. and GÖRÖG, S.: *Acta Chim. Hung.* **22**, 159 (1960).
8. SAJÓ, I.: *Acta Chim. Hung.* **16**, 115 (1958).
9. PECSOK, R. L. and MAVERICK, E. F.: *J. Amer. Chem. Soc.* **76**, 358 (1954).
10. BUSEV, A. I., TIPDIVA, V. G. and SOKOLOVA, T. A.: *Zh. Neorg. Khim.* **5**, 2749 (1960).
11. SCHWARZENBACH, G. and SANDERA, J.: *Helv. Chim. Acta* **36**, 1089 (1953).
12. RINGBOM, A., SÜTONEN, S. and SKRIFVARS, B.: *Acta Chem. Scand.* **11**, 551 (1957).
13. MORGEN, L. O. and JUSTUS, N. L.: *J. Amer. Chem. Soc.* **78**, 38 (1956).

TABLE 41. STABILITY CONSTANTS OF ETHYLENEDIAMINE COMPLEXES

Central atom	$\log K_1$	$\log K_2$	$\log K_3$	Measuring method	Ionic strength of medium	Temperature, °C	Refs.
H^+	10.09			pH-metric	→0	20	1, 2
H^+	10.18			pH-metric	1 KNO_3	25	3
Cd^{2+}	5.63	4.59	2.07	pH-metric	1 KNO_3	25	3
Co^{2+}	5.93	4.73	3.30	pH-metric	1 KCl	25	4
Cu^{2+}	10.72	9.31		pH-metric	1 KNO_3	25	5
Fe^{2+}	4.34	3.31	2.05	pH-metric	1.4	25	6
Mn^{2+}	2.77	2.10	0.92	pH-metric	1.4	25	6
Ni^{2+}	7.53	6.32	4.49	pH-metric	→0	20	1, 2
Zn^{2+}	5.77	5.06	3.28	pH-metric	→0	20	1, 2

1. BERTSCH, C. R., FERNELIUS, W. C. and BLOCK, B. P.: *J. Phys. Chem.* **62**, 444 (1958).
2. McINTYRE, G. H., BLOCK, B. P. and FERNELIUS, W. C.: *J. Amer. Chem. Soc.* **81**, 529 (1959).
3. BJERRUM, J. and ANDERSON, P.: *Kgl. Danske Videnskab. Selskab* **7**, 22 (1945).
4. EDWARDS, L. J.: Dissertation, Univ. Michigan, 1950.
5. BJERRUM, J. and NELSON, E. J.: *J. Acta Chem. Scand.* **2**, 307 (1948).
6. PECZOK, P. L. and BJERRUM, J.: *Acta Chem. Scand.* **11**, 1419 (1957).

TABLE 42. STABILITY CONSTANTS OF ETHYLXANTHATE COMPLEXES

Central atom	$\log \beta_2$	$\log \beta_3$	Measuring method	Ionic strength of medium	Temperature, °C	Ref.
Co^{2+}	7.20	11.25	photometric	Water–methanol 1:3 mixture 0.3 M $NaClO_4$	25	1
Cu^{2+}	9.56		photometric	Water–methanol 1:3 mixture 0.3 M $NaClO_4$	25	1
Ni^{2+}	7.74		photometric	Water–methanol 1:3 mixture 0.3 M $NaClO_4$	25	1

1. BURGER, K., PAPP-MOLNÁR, E., VÁSÁRHELYI-NAGY, H. and KORECZ, L.: *Acta Chim. Hung.* **64**, 323 (1970).

TABLE 43. STABILITY CONSTANTS OF 8-HYDROXY-2-METHYLQUINOLINE COMPLEXES

Central atom	$\log K_1$	$\log K_2$	Measuring method	Ionic strength of medium	Temperature, °C	Refs.
H^+	11.01		pH-metric	50% dioxane, 0.3 $NaClO_4$	20	1
H^+	11.45		pH-metric	50% dioxane, 0.3 $NaClO_4$	25	2
Cu^{2+}	10.22	9.32	pH-metric	50% dioxane, 0.3 $NaClO_4$	20	1
Fe^{2+}	8.75	8.35	pH-metric	50% dioxane, 0.3 $NaClO_4$	25	2
Mg^{2+}	3.73	3.13	pH-metric	50% dioxane, 0.3 $NaClO_4$	20	1
Ni^{2+}	8.52	7.96	pH-metric	50% dioxane, 0.3 $NaClO_4$	20	1
UO_2^{2+}	~9.4	~8	pH-metric	50% dioxane, 0.3 $NaClO_4$	20	1
Zn^{2+}	8.66	8.10	pH-metric	50% dioxane, 0.3 $NaClO_4$	20	1

1. IRVING, H. and ROSSOTTI, H. S.: *J. Chem. Soc.* **1954**, 2910.
2. TOMKINSON, J. C. and WILLIAMS, R. J. P.: *J. Chem. Soc.* **1958**, 1153.

TABLE 44. STABILITY CONSTANTS OF 8-HYDROXYQUINOLINE COMPLEXES

Central atom	$\log K_1$	$\log K_2$	$\log K_3$	$\log K_4$	Measuring method	Ionic strength of medium	Temperature, °C	Refs.
H^+	10.80				pH-metric	50% dioxane, 0.3 $NaClO_4$	20	1
H^+	11.54				pH-metric	50% dioxane	25	2
Cd^{2+}	9.43	7.68			pH-metric	50% dioxane	25	2
Ce^{3+}	9.15	7.98			pH-metric	50% dioxane	25	2
Co^{2+}	10.55	9.11			pH-metric	50% dioxane	25	2
Cu^{2+}	13.03	12.35			pH-metric	50% dioxane, 0.3 $NaClO_4$	20	1
Fe^{2+}	8.71	8.12	5.30		pH-metric	50% dioxane, 0.3 $NaClO_4$	25	3
Fe^{3+}			$\log \beta_3$ 38.00		pot. redox electr.	50% dioxane, 0.3 $NaClO_4$	25	3
La^{3+}	8.66	7.74			pH-metric, sol. measmnt.	50% dioxane	25	4
Mg^{2+}	5.04	4.29			pH-metric	50% dioxane, 0.3 $NaClO_4$	20	5
Mn^{2+}	8.28	7.17			pH-metric	50% dioxane	25	2
Ni^{2+}	10.43	9.97			pH-metric	50% dioxane, 0.3 $NaClO_4$	20	1
Th^{4+}	10.45	9.95	9.45	8.95	extr.	0.1	25	6
UO_2^{2+}	11.25	9.64			pH-metric	50% dioxane	20	1
Zn^{2+}	9.34	8.22			pH-metric	50% dioxane, 0.3 $NaClO_4$	20	1

1. IRVING, H. and ROSSOTTI, H. S.: *J. Chem. Soc.* **1954**, 2910.
2. JOHNSTON, W. D. and FREISER, H.: *J. Amer. Chem. Soc.* **74**, 5239 (1952).
3. TOMKINSON, J. C. and WILLIAMS, R. J. P.: *J. Chem. Soc.* **1958**, 2010.
4. JOHNSTON, W. D. and FREISER, H.: *Anal. Chim. Acta* **11**, 201 (1954).
5. NÄSÄNEN, R.: *Suomen Kem.* **26**B, 2, 11 (1953).
6. DYRSSEN, D.: *Svenska. Kem. Tidok.* **65**, 43 (1953).

TABLE 45. STABILITY CONSTANTS OF 8-HYDROXYQUINOLINE-5-SULFONIC ACID COMPLEXES

Central atom	$\log K_1$	$\log K_2$	$\log K_3$	$\log K_4$	Measuring method	Ionic strength of medium	Temperature, °C	Refs.
H^+	8.76				pH-metric	→0	25	1
H^+	8.53				pH-metric	→0	25	2
Ba^{2+}	2.31				photometr. +pH-metr.	→0	25	1
Ca^{2+}	3.52				photometr. +pH-metr.	→0	25	1
Cd^{2+}	7.70	6.5			photometr. +pH-metr.	→0	25	1
Ce^{3+}	6.05	5.0	3.9		pH-metric	→0	25	2
Co^{2+}	8.82	7.1			photometr. +pH-metr.	→0	25	1
Cu^{2+}	11.53	10.1			photometr. +pH-metr.	→0	25	1
Er^{3+}	7.16	6.18	5.22		pH-metric	→0	25	2
Fe^{2+}		$\log \beta_2$ 15.7	$\log \beta_3$ 21.7		pH-metric	0.3 NaCl	25	3
Gd^{3+}	6.64	5.73	4.9		pH-metric	→0	25	2
La^{3+}	5.63	4.50	3.70		pH-metric	→0	25	2
Mg^{2+}	4.79	3.04			pH-metric	→0	25	1
Mn^{2+}	6.94				pH-metric	→0	25	1
Nd^{3+}	6.3	5.3	4.4		pH-metric	→0	25	2
Ni^{2+}	9.57	8.7			photometr. +pH-metr.	→0	25	1
Pb^{2+}	8.53	7.6			pH-metric	→0	25	1
Pr^{3+}	6.17	5.20	4.3		pH-metric	→0	25	2
Sr^{2+}	2.75				photometr. +pH-metr.	→0	25	1
Th^{4+}	9.56	8.73	7.62	6.12	pH-metric	0.1 KNO_3	25	4
UO_2^{2+}	8.52	7.16			pH-metric	→0	25	4
Zn^{2+}	8.65	7.5			photometr. +pH-metr.	→0	25	4

1. NÄSÄNEN, R. and UISITALO, E.: *Acta Chem. Scand.* **8,** 112, 835 (1954).
2. FREASIER, B. F. OBERG, A. G. and WENDLANDT, W. W.: *J. Phys. Chem.* **62,** 700 (1958).
3. TOMKINSON, J. C. and WILLIAMS, R. J. P.: *J. Chem. Soc.* **1958,** 1153, 2010.
4. RICHARD, C. F., GUSTAFSON, R. L. and MARTELL, A. E.: *J. Amer. Chem. Soc.* **81,** 1033 (1959).

TABLE 46. STABILITY CONSTANTS OF MERCAPTOACETIC ACID COMPLEXES

Central atom	log K_1	log β_2	Measuring method	Ionic strength of medium	Temperature, °C	Refs.
H^+	10.20		pH-metric	0.1 KCl	25	1
Co^{2+}	5.84	12.15	pH-metric	0.1	25	1
Fe^{2+}		10.92	sol. measmnt.	→0	25	2
Hg^{2+}		43.82	potentiometric	1 KNO_3	25	3
Mn^{2+}	4.38	7.56	pH-metric	0.1	25	1
Ni^{2+}	6.98	13.53	pH-metric	0.1	25	1
Zn^{2+}	7.86	15.04	pH-metric	0.1	25	1

1. LEUSSING, L. D.: *J. Amer. Chem. Soc.* **80**, 4810 (1958).
2. LEUSSING, L. D. and KOLTHOFF, I. M.: *J. Amer. Chem. Soc.* **75**, 3904 (1953).
3. STRICKS, W., KOLTHOFF, I. M. and HEYNDRICK, A.: *J. Amer. Chem. Soc.* **76**, 1515 (1954).

TABLE 47. STABILITY CONSTANTS OF NIOXIME COMPLEXES

Central atom	log K_1	log β_2	Measuring method	Ionic strength of medium	Temperature, °C	Refs.
H^+	12.0		pH-metric	50% dioxane	25	1
H^+	13.11		pH-metric	75% dioxane	25	2
Co^{2+}		25.5	pH-metric	50% dioxane	25	1
Cu^{2+}		25.7	pH-metric	50% dioxane	25	1
Mn^{2+}	8.2	15.4	pH-metric	50% dioxane	25	3
Ni^{2+}	11.09	22.46	pH-metric	75% dioxane	25	2
Zn^{2+}		14.9	pH-metric	50% dioxane	25	3

1. PESHKOVA, V. M. and BOCHKOVA, V. M.: *Nauchn. Dokl. Vis. Shkoly, Khim. i Khim. Tekhnol.* **1**, 62 (1958).
2. BANKS, C. V. and ANDERSON, S.: *Inorg. Chem.* **2**, 112 (1963).
3. BURGER, K., RUFF, I. and PAPP-MOLNÁR, E.: *Annales Univ. Sci. Budapest* **7**, 49 (1965).

Table 48. Stability Constants of Nitrilotriacetate Complexes

Central atom	$\log K_1$	$\log K_2$	Measuring method	Ionic strength of medium	Temperature, °C	Refs.
H^+	9.75		pH-metric	0.1 KNO_3	25	1
Ba^{2+}	4.72		pH-metric	0.1 KNO_3	25	2
Ca^{2+}	6.57		pH-metric	0.1 KNO_3	25	1
Cd^{2+}	9.54		pH-metric	0.1 KCl	20	3
Ce^{3+}	10.83		pH-metric	0.1 KNO_3	25	1
Co^{2+}	10.6		pH-metric	0.1 KCl	20	3
Cu^{2+}	13.10		pH-metric	0.1 KNO_3	25	1
Dy^{3+}	11.74	9.41	pH-metric	0.1 KNO_3	25	1
Er^{3+}	12.03	9.26	pH-metric	0.1 KNO_3	25	1
Eu^{3+}	11.52	9.18	pH-metric	0.1 KNO_3	25	1
Fe^{2+}	8.83		pH-metric	0.1 KCl	20	4
Fe^{3+}	15.87	8.45	pot. redox electr.	0.1 KCl	20	5
Gd^{3+}	11.54	9.26	pH-metric	0.1 KNO_3	25	1
Ho^{3+}	11.90	9.35	pH-metric	0.1 KNO_3	25	1
La^{3+}	10.36	7.24	pH-metric	0.1 KNO_3	25	1
Lu^{3+}	12.49	9.42	pH-metric	0.1 KNO_3	25	1
Mg^{2+}	5.36		pH-metric	0.1 KNO_3	25	2
Mn^{2+}	7.44		pot. H electr.	0.1 KCl	20	3
Nd^{3+}	11.26	8.47	pH-metric	0.1 KNO_3	25	1
Ni^{2+}	11.26		pH-metric	0.1 KCl	20	3
Pb^{2+}	11.8		pH-metric	0.1 KCl	20	3
Pr^{3+}	11.07	8.18	pH-metric	0.1 KNO_3	25	1
Sm^{3+}	11.53	9.00	pH-metric	0.1 KNO_3	25	1
Sr^{2+}	4.91		pH-metric	0.1 KNO_3	25	2
Tb^{3+}	11.59	9.38	pH-metric	0.1 KNO_3	25	1
Th^{4+}	12.4		pH-metric	0.1 KNO_3	25	6
Tm^{3+}	12.25	9.23	pH-metric	0.1 KNO_3	25	1
Y^{3+}	11.48	8.95	pH-metric	0.1 KNO_3	25	1
Zn^{2+}	10.45		pH-metric	0.1 KCl	20	3
Zr^{4+}	20.8		photometric	0.1	25	7

1. Moeller, T. and Ferrus, R.: *Inorg. Chem.* **1**, 55 (1962).
2. Bohigian, T. A. and Martell, A. E.: Progr. Rep. US Atom. En. Comm. Contr. No AT (30–1)–1823 (1960).
3. Schwarzenbach, G. and Freitag, E.: *Helv. Chim. Acta* **34**, 1492 (1951).
4. Schwarzenbach, G., Anderegg, G., Schneider, W. and Senn, H.: *Helv. Chim. Acta* **38**, 1147 (1955).
5. Schwarzenbach, G. and Heller, J.: *Helv. Chim. Acta* **34**, 1889 (1951).
6. Courtney, R. C., Gustafson, R. L., Chaberek, S. and Martell, A. E.: *J. Amer. Chem. Soc.* **80**, 2121 (1958).
7. Intorre, B. J. and Martell, A. E.: *Inorg. Chem.* **3**, 81 (1964).

TABLE 49. STABILITY CONSTANTS OF 1,10-PHENANTHROLINE COMPLEXES

Central atom	$\log K_1$	$\log K_2$	$\log K_3$	Measuring method	Ionic strength of medium	Temperature, °C	Refs.
H^+	4.95			pH-metric	0.1 $NaNO_3$	20	1
Ag^+	5.02	7.05		potentiometric Ag electr.	0.1	25	2
Ca^+	0.7			pH-metric	0.1 $NaNO_3$	20	1
Cd^{2+}	5.78	5.04	4.10	pH-metric	0.1 $NaNO_3$	20	1
Co^{2+}	7.25	6.70	5.95	pH-metric	0.1 $NaNO_3$	20	1
Cu^{2+}	9.25	6.75	5.35	pot. Hg electr.	0.1 $NaNO_3$	20	1
Fe^{2+}	$K_1 K_2 < K_3$		$\log \beta_3$ 21.3	photometric	0.1 $NaNO_3$	20	1
Fe^{3+}	6.5	4.9	12.12	potentiometric			3
Hg^{2+}		$\log \beta_2$ 19.65	3.7	potentiometric Hg electr.	0.1 $NaNO_3$	20	1
Mg^{2+}	1.2			PH-metric	0.1 KNO_3	20	1
Mn^{2+}	4.13	3.48	2.7	pH-metric	0.1 KNO_3	20	1
Ni^{2+}	8.8	8.3	7.7	pot. Hg electr.	0.1 $NaNO_3$	20	1
Pb^{2+}	4.65			pH-metric	0.1 $NaNO_3$	20	1
VO^{2+}	5.47	4.22		pH-metric	0.082	25	4
Zn^{2+}	6.55	5.80	5.20	pH-metric	0.1 $NaNO_3$	20	1

1. ANDEREGG, G.: *Helv. Chim. Acta* **46**, 2397 (1963).
2. DALE, J. M. and BANKS, C. V.: *Inorg. Chem.* **2**, 591 (1963).
3. ANDEREGG, G.: *Helv. Chim. Acta* **45**, 1643 (1962).
4. TRUJILLO, R. and BRITO, F.: *Anal. Real Soc. Esp. Fis. Quim.* **53B**, 533 (1957).

TABLE 50. STABILITY CONSTANTS OF PYRIDINE-2-ALDOXIME COMPLEXES

Central atom	$\log K_1$	$\log K_2$	$\log K_3$	Measuring method	Ionic strength of medium	Temperature, °C	Refs.
H^+	3.4	10.0		pH-metric	0.3 $NaClO_4$	25	1
Co^{2+}	8.8	8.8		photometric	0.3 $NaClO_4$	25	1
Cu^{2+}	8.9	5.65		pH-metric			2
Fe^{2+}	$K_1K_2 < K_3$		$\log \beta_3$ 24.85	photometric	0.045	25	3
Mn^{2+}	5.2	3.9		pH-metric	0.3 $NaClO_4$	25	1
Ni^{2+}	9.4	7.1	5.5	pH-metric	0.3 $NaClO_4$	25	1
Zn^{2+}	5.8	5.3		pH-metric	0.3 $NaClO_4$	25	1

1. BURGER, K., EGYED, I. and RUFF, I.: *J. Inorg. Nucl. Chem.* **28,** 139 (1966).
2. KIRSON, B.: *Bull. Soc. Chim. France* **1962,** 1030.
3. HANANIA, G. I. H. and IRVINE, D. H.: *J. Chem. Soc.* **1962,** 2745.

TABLE 51. STABILITY CONSTANTS OF PYROLIDINE DITHIOCARBAMATE COMPLEXES

Central atom	$\log \beta_2$	$\log \beta_3$	Measuring method	Ionic strength of medium	Temperature, °C	Ref.
Co^{2+}		15.90	photometric	Water–methanol 1:3 mixture 0.3 M $NaClO_4$		1
Fe^{2+}		12.69	photometric	Water–methanol 1:3 mixture 0.3 M $NaClO_4$		1
Ni^{2+}	11.0		photometric	Water–methanol 1:3 mixture 0.3 M $NaClO_4$		1

1. PAPP-MOLNÁR, E., VÁSÁRHELYI-NAGY, H. and BURGER, K.: *Acta Chim. Hung.* **64,** 317 (1970).

TABLE 52. STABILITY CONSTANTS OF SALICYLALDOXIME COMPLEXES

Central atom	$\log K_1$	$\log K_2$	Measuring method	Ionic strength of medium	Temperature, °C	Ref.
H^+	10.7		pH-metric	75% dioxane, 0.1 $NaClO_4$	20	1
Co^{2+}	6.4	7.1	pH-metric	75% dioxane, 0.1 $NaClO_4$	20	1
Cu^{2+}	$K_1 \ll K_2$	$\log \beta_2$ 21.5	photometric	75% dioxane, 0.1 $NaClO_4$	20	1
Fe^{2+}	9.38	7.35	pH-metric	75% dioxane, 0.1 $NaClO_4$	20	1
Mn^{2+}	5.8	6.1	pH-metric	75% dioxane, 0.1 $NaClO_4$	20	1
Ni^{2+}	6.9	7.4	pH-metric	75% dioxane, 0.1 $NaClO_4$	20	1
Zn^{2+}	6.3	7.2	pH-metric	75% dioxane, 0.1 $NaClO_4$	20	1

1. BURGER, K. and EGYED, I.: *J. Inorg. Nucl. Chem.* **27**, 2361 (1965).

TABLE 53. STABILITY CONSTANTS OF SULFOSALICYLIC ACID COMPLEXES

Central atom	$\log K_1$	$\log K_2$	$\log K_3$	Measuring method	Ionic strength of medium	Temperature, °C	Refs.
H^+	12.00			pH-metric	0.1 $NaClO_4$	25	1
H^+	11.74			pH-metric	3.0 $NaClO_4$	25	2
Al^{3+}	13.20	9.63	6.06	pH-metric	0.1 $NaClO_4$	25	1
Be^{2+}	11.71	9.10		pH-metric	0.1 $NaClO_4$	25	1
Co^{2+}	6.13	3.69		pH-metric	0.1 $NaClO_4$	25	1
Cr^{3+}	9.56			pH-metric	0.1 $NaClO_4$	25	1
Cu^{2+}	9.52	6.93		pH-metric	0.1 $NaClO_4$	25	1
Fe^{3+}	14.42	10.76	7.06	pH-metric	3.0 $NaClO_4$	25	2
Mn^{2+}	5.24	3.00		pH-metric	0.1 $NaClO_4$	25	1
Ni^{2+}	6.42	3.82		pH-metric	0.1 $NaClO_4$	25	1
UO_2^{2+}	11.14	8.06		pH-metric	0.1 $NaClO_4$	25	1
Zn^{2+}	6.05	4.6		pH-metric	0.1–0.15 KCl	20	3

1. BANKS, C. V. and SINGH, R. S.: *J. Amer. Chem. Soc.* **81**, 6159 (1960).
2. AGREN, A.: *Acta Chem. Scand.* **8**, 266 (1954).
3. PERRIN, D. D.: *Nature* **182**, 1958.

TABLE 54. STABILITY CONSTANTS OF ($\pm$)-TARTARIC ACID COMPLEXES

Central atom	$\log K_1$	$\log K_2$	Measuring method	Ionic strength of medium	Temperature, °C	Refs.
H^+	4.366		potentiometric, H electr. Ag/AgCl electr.	$\rightarrow 0$	25	1
H^+	16.5		pH-metric	75% dioxane	30	2
Ba^{2+}	1.68		ion exchange	0.16	25	3
Ca^{2+}	1.78		ion exchange	0.16	25	6
Cu^{2+}	3.00	2.11	potentiometric	1 $NaClO_4$	20	4
Fe^{3+}	7.49		photometric			5
Mg^{2+}	7.9	5.3	pH-metric	75% dioxane	30	2
Ni^{2+}	9.9	6.9	pH-metric	75% dioxane	30	2
Sr^{2+}	1.59		ion exchange	0.16	25	6

1. BATES, R. G. and CANHAM, A. G.: *J. Res. Nat. B. St.* **47**, 5 (1951).
2. VAN UITERT, L. G. and FERNELIUS, W. C.: *J. Amer. Chem. Soc.* **76**, 375 (1954).
3. SCHUBERT, J.: *J. Amer. Chem. Soc.* **76**, 3442 (1954).
4. FRONEAUS, S.: Dissertation, London 1948.
5. PYATNICKY, I. V. and GENDLER, S. M.: *Zh. Obshch. Khim.* **26**, 2137 (1956).
6. SCHUBERT, J. and LINDENBAUM, A.: *J. Amer. Chem. Soc.* **74**, 3529 (1952).

TABLE 55. STABILITY CONSTANTS OF THENOYLTRIFLUOROACETONE COMPLEXES

Central atom	$\log K_1$	$\log K_2$	Measuring method	Ionic strength of medium	Temperature, °C	Refs.
H^+	5.70		pH-metric	0.1	20	1
H^+	6.10		photometric	$\rightarrow 0$	25	2
Cu^{2+}	6.55	13.0	pH-metric	0.1	20	1
Fe^{3+}	6.9		photometric	$\rightarrow 0$	25	2
Ni^{2+}		10.0	pH-metric	0.1	20	1
Pr^{3+}	9.53		extr.	0.1 NH_4Cl	25	3
Pu^{3+}	9.53		extr.	0.1 NH_4Cl	25	3
Pu^{4+}	8.0		photometric	$\rightarrow 0$	25	2
Th^{4+}	8.1		photometric	$\rightarrow 0$	25	2
U^{4+}	7.2		photometric	~ 0.1	25	2

1. VAN UITERT, L. G., FERNELIUS, W. C. and DOUGLAS, B. E.: AEC Rept. NYO- 626, March 1, 1951.
2. Calvin's data (SILLEN, L. G. and MARTELL, A. E.: *Stability Constants*).
3. KEENAN, T. K. and SUTTLE, J. F.: *J. Amer. Chem. Soc.* **76,** 2184 (1954).

TABLE 56. STABILITY CONSTANTS OF THIONALID COMPLEXES

Central atom	$\log K_1$	$\log K_2$	$\log K_3$	Measuring method	Ionic strength of medium	Temperature, °C	Ref.
H^+	10.2			pH-metric	75% dioxane, 0.1 $NaClO_4$	20	1
Co^{2+}	7.3	6.8	6.0	pH-metric	75% dioxane, 0.1 $NaClO_4$	20	1
Cu^{2+}	10.3	9.7	8.4	pH-metric	75% dioxane, 0.1 $NaClO_4$	20	1
Fe^{2+}	7.0	6.7	5.5	pH-metric	75% dioxane, 0.1 $NaClO_4$	20	1
Mn^{2+}	4.4	4.4		pH-metric	75% dioxane, 0.1 $NaClO_4$	20	1
Ni^{2+}	6.9	6.5		pH-metric	75% dioxane, 0.1 $NaClO_4$	20	1
Zn^{2+}	7.8	7.1		pH-metric	75% dioxane, 0.1 $NaClO_4$	20	1

1. BURGER, K., KORECZ, L. and TÓTH, A.: *Magy. Kém. Foly.* **73,** 455 (1967).

TABLE 57. STABILITY CONSTANTS OF THIOUREA COMPLEXES

Central atom	$\log K_1$	$\log K_2$	$\log \beta_3$	$\log \beta_4$	Measuring method	Ionic strength of medium	Temperature, °C	Refs.
H^+	2.03				pH-metric	0.01	25	1
Ag^+			13.05		Ag electrode	$\to 0$	25	2
Cd^{2+}	1.38	1.71	1.60	3.55	polarographic	0.1 KNO_3	25	3
Pb^{2+}	0.60	1.04	0.98	2.04	polarographic	0.1 KNO_3	25	3

1. LANE, T. J., RYAN, J. A. and WALTER, J. L.: *J. Amer. Chem. Soc.* **78**, 5560 (1956).
2. FYFE, W. S.: *J. Chem. Soc.* **1955**, 1032.
3. LANE, T. J., RYAN, J. A. and BRITTEN, E. F.: *J. Amer. Chem. Soc.* **80**, 315 (1958).

TABLE 58. STABILITY CONSTANTS OF TIRON COMPLEXES

Central atom	$\log K_1$	$\log K_2$	$\log K_3$	Measuring method	Ionic strength of medium	Temperature, °C	Refs.
H^+	11.6			pH-metric	1 $NaClO_4$	25	1
Al^{3+}	19.02	12.08	2.4	pH-metric		25	2, 3
Cd^{2+}	7.69	5.0		pH-metric	1 $NaClO_4$	25	1
Co^{2+}	8.19	6.22		pH-metric	1 $NaClO_4$	25	1
Cu^{2+}	12.76	10.97		pH-metric	1 $NaClO_4$	25	1
Fe^{3+}	20.7	15.2	11.0	pH-metric	0.1 KCl	20	4
Ni^{2+}	8.56	6.34		pH-metric	1 $NaClO_4$	25	1
UO_2^{2+}	15.90			pH-metric	0.1 KNO_3	25	5
VO^{2+}	15.88			pH-metric	0.1	25	6
Zn^{2+}	9.00	7.91		pH-metric	1 $NaClO_4$	25	1

1. NÄSÄNEN, R.: *Suomen Kem.* **33**, 7, 11 (1960).
2. NÄSÄNEN, R.: *Suomen Kem.* **30B**, 61 (1957).
3. NÄSÄNEN, R. and VEIVO, J.: *Suomen Kem.* **29B**, 213 (1956).
4. WILLI, A. and SCHWARZENBACH, G.: *Helv. Chim. Acta* **34**, 528 (1951).
5. GUSTAFSON, R. L., RICHARD, C. and MARTELL, A. E.: Progress Rept. 1958 U.S. Atom. En. Comm. No AT(30-1)-1823.
6. GUSTAFSON, R. L., RICHARD, C. and MARTELL, A. E.: *J. Amer. Chem. Soc.* **82**, 1526 (1960).

SOME MORE IMPORTANT ORGANIC SOLVENTS

Solvents constitute a special group of organic reagents. In the analytical chemistry of metals the significance of solvents is connected primarily with solvent extraction procedures. The proper choice of the solvent is a requirement of the successful accomplishment of solvent extraction procedures. However, even with a properly chosen solvent, the analyst must take into account the possible errors caused by various contaminations of the solvent. Thus, for instance, the peroxide contaminants of ethers and certain other oxygen-containing solvents may oxidize some ligands and thus diminish the concentration of excess reagent which may be necessary for the formation of the complex. In some cases the metal ion or the already formed complex may undergo oxidation.

Organic solvents miscible with water are frequently used for the enhancement of the solubility of certain complexes in water. Thus there are complexes which are insoluble in water, but dissolve in water–organic solvent mixture, and hence these are suitable for the spectrophotometric determination of the metal concerned.

In such examinations the exactly constant composition of the solvent system is an important requirement of successful work. The moisture content of the solvent, for instance, is a possible source of error, particularly with solvents miscible with water.

Table 59 contains in addition to the density and boiling point data, which are most useful in the identification and quality control of solvents, the dielectric constants, Gutmann's donor numbers (donicity) and the solubilities in water too, which are of value in the choice of the analytically most appropriate solvent.

TABLE 59. SOME IMPORTANT ORGANIC SOLVENTS

Solvent	Formula	Density, (°C)	Boiling point	Di-electric constant	Donor number	Solubility in water, (°C)
Acetic anhydride	$(CH_3CO)_2O$	1.0811 (20)	140.0	22.1	10.5	miscible
Acetic acid	CH_3COOH	1.049 (20)	118.1	6.1		miscible
Acetone	CH_3COCH_3	0.7908 (20)	56.5	20.7	17.0	miscible
Acetonitrile	CH_3CN	0.7767	81.6	38.8	14.1	miscible
Acetylacetone	$CH_3COCH_2COCH_3$	0.9753	140.5	25.7		dissolves in water containing HCl
n-Amyl acetate	$CH_3COOC_5H_{11}$	0.8753 (20)	149.2	4.8		0.2 v/v%
n-Amyl alcohol	$CH_3(CH_2)_3CH_2OH$	0.8144 (20)	138.06	13.9		2.19 g/g%
Benzene	C_6H_6	0.8944 (0)	80.103	2.3		0.180 g/100 g (25)
Benzyl alcohol	$C_6H_5CH_2OH$	1.05 (15)	205.2	13.1		4% (17)
1-Butanol	$CH_3(CH_2)_3OH$	0.8134 (15)	117.71	17.1		79 g/l. (20)
n-Butyl acetate	$CH_3COOC_4H_9$	0.8813 (20)	126.5	5.0		0.5% (25)
n-Butyl ether	$CH_3(CH_2)_3O(CH_2)_3CH_3$	0.769 (20)	142	3.1		practically insoluble
Carbon disulfide	CS_2	1.2626 (20)	46.3	2.6		2.2 g/l. (22)
Carbon tetrachloride	CCl_4	1.595 (20)	76–77	2.2		0.8 g/l. (20)
Chloroform	$CHCl_3$	1.4985 (15)	61.26	4.8		10 g/l. (15)
Cyclohexane	C_6H_{12}	0.7831 (15)	80.738	2.0		0.01 g/100 g (20)
Cyclohexanol	$C_6H_{11}OH$	0.9684 (25)	161.5	15.0		0.567% (15)

Cyclohexanone	$(CH_2)_5CO$	0.951 (15)	156.7	18.3	5 g/100 ml (30)
Dibutylamine	$(C_4H_9)_2NH$	0.7601 (20)	159-161		soluble
m-Dichlorobenzene	$C_6H_4Cl_2$	1.2828 (25)	173.0	5.0	0.0123 g/100 ml (25)
o-Dichlorobenzene	$C_6H_4Cl_2$	1.3003 (25)	180.48	9.9	practically insoluble
p-Dichlorobenzene	$C_6H_4Cl_2$	1.4581 (20.5)	174.12	2.4	0.077 g/100 g (30)
Dichloroethane	CH_3CHCl_2	1.2383 (30)	83.48	10.6	0.1
Dichloromethane	CH_2Cl_2	1.336 (20)	40.1	9.1	20 g/l. (20)
Diethanolamine	$(HOCH_2CH_2)_2NH$	1.0966 (20)	269.1		soluble
Diethylamine	$(C_2H_5)_2NH$	0.7108 (18)	55.5	3.6	soluble
Diethyleneglycol	$HO(CH_2)_2O(CH_2)_2OH$	1.177	244.5		soluble
Diisobutyl ketone	$(CH_3)_2CHCH_2COCH_2CH(CH_3)_2$	0.938	164-166		miscible
Diisopropyl ketone	$(CH_3)_2CHCOCH(CH_3)_2$	0.8062 (20)	123.7		practically insoluble
Dimethylformamide	$(CH_3)_2CNH$	0.9443 (25)	153.0	36.71	27.0	miscible
Dimethylsulphoxide	$(CH_3)_2SO$	1.100 (20)	189.0	48.9	29.8	miscible
p-Dioxan	$C_4H_8O_2$	1.03375 (20)	101.32	2.2		miscible
Dipropylamine	$(C_3H_7)_2NH$	0.7340 (20)	110.7	2.9	soluble
Ethanol	C_2H_5OH	0.7893 (20)	78.325	24.3	miscible
Ethyl acetate	$CH_3COOC_2H_5$	0.901 (20)	77.15	6.0	17.1	8.6% (20)
Ethyleneglycol	$HOCH_2CH_2OH$	1.1171 (15)	197.2	37.7	miscible
Ethyl ether	$C_2H_5OC_2H_5$	0.7193 (15)	34.5	4.3	19.2	7.42 g/g % (20)
Furfuryl alcohol	$C_4H_3OCH_2OH$	1.1238 (30)	170		miscible

TABLE 59 (*continued*)

Solvent	Formula	Density, (°C)	Boiling point	Di-electric constant	Donor number	Solubility in water, (°C)
Glycerol	CH$_2$OHCHOHCH$_2$OH	1.2613 (20)	290	42.5		miscible
Heptane	CH$_3$(CH$_2$)$_5$CH$_3$	0.684 (20)	98.52	1.9		0.052 g/l. (15.5)
2-Heptanone	CH$_3$CO(CH$_2$)$_4$CH$_3$	0.822 (15)	150	11.9		slightly soluble
Hexamethylphosphoramide	(CH$_3$)$_6$N$_3$PO	1.02	230 (739 torr)	33.5	38.8	miscible
Hexane	CH$_3$(CH$_2$)$_4$CH$_3$	0.6603 (20)	69.0	1.9		0.138 g/l. (15.5)
Isoamyl alcohol	(CH$_3$)$_2$CHCH$_2$CH$_2$OH	0.8129 (15)	130.5	14.7		2.67 g/g%
Isobutyl acetate	CH$_3$COOCH$_2$CH(CH$_3$)$_2$	0.871 (20)	116.5	5.3		0.63% (25)
Isobutyl alcohol	(CH$_3$)$_2$CHCH$_2$OH	0.8169 (20)	107–108	17.7		95 g/l.
Isopropyl acetate	CH$_3$COOCH$_2$(CH$_3$)$_2$	0.869 (25)	89	18.3		3.09% (20)
Isopropyl alcohol	(CH$_3$)$_2$CHOH	0.7892 (15)	82.3	18.3		miscible
Isopropyl ether	(CH$_3$)$_2$CHOCH(CH$_3$)$_2$	0.7281 (20)	67.5	3.9		0.65 v/v% (25)
Mesitylene	(CH$_3$)$_3$C$_6$H$_3$	0.8634 (20)	164.6	2.3		insoluble
Mesityloxide	(CH$_3$)$_2$C=CHCOCH$_3$	0.8539 (20)	128.7	15.6		3 g/100 ml
Methanol	CH$_3$OH	0.7961 (15)	64.7	32.6		miscible
Methyl acetate	CH$_3$COOCH$_3$	0.9274 (25)	57.1	6.7	16.5	31.9% (20)
Methyl ethyl ketone	CH$_3$CH$_2$COCH$_3$	0.805 (20)	79.6	18.5		35.3% (10)

Name	Formula					
Methyl isobutyl carbinol	$(CH_3)_2CHCH_2CHOHCH_3$	0.8075 (25)	131.4			18 g/l.
Methyl isobutyl ketone (hexanone)	$(CH_3)_2CHCH_2COCH_3$	0.8006 (20)	115.8	13.1		2% (20)
Nitrobenzene	$C_6H_5NO_2$	1.193 (25)	210.8	34.8	4.4	9.5 ml/100 ml
Nitromethane	CH_3NO_2	1.1448 (15)	101.25	35.9	2.7	soluble
2-Pentanone	$CH_3CO(CH_2)_2CH_3$	0.812 (15)	101.7	15.4		
3-Pentanone	$(C_2H_5)_2CO$	0.8095 (25)	101.7	17.0		4.7 g/100 ml (20)
1-Propanol	$CH_3(CH_2)_2OH$	0.8075 (15)	97.2	20.1		miscible
n-Propyl acetate	$CH_3COOC_3H_7$	0.8867 (20)	101.6	5.7		1.89% (20)
Propyleneglycol	$CH_3CHOHCH_2OH$	1.0364 (20)	189	32.0		miscible
n-Propyl ether	$CH_3(CH_2)_2O(CH_2)_2CH_3$	0.7518 (15)	91	3.4		0.25 g/g%
Pyridine	C_5H_5N	0.9878 (15)	115.3	12.3	33.1	soluble
Quinoline	C_9H_7N	1.095 (20)	237.7	9.0		6%
Tetrachloroethylene	C_2Cl_4	1.6311 (15)	121.20	2.3		0.015 g/100 g
Toluene	$C_6H_5CH_3$	0.866 (20)	110.8	2.4	23.7	0.47 g/l. (16)
Tributyl phosphate	$(C_4H_9)_3PO_4$	0.9727 (27)	177–178	8.0		0.6%
Trichloroethylene	C_2HCl_3	1.4556 (25)	87	3.4		1 g/l.
m-Xylene	$C_6H_4(CH_3)_2$	0.8684 (15)	138.8	2.4		0.196 g/l. (25)
o-Xylene	$C_6H_4(CH_3)_2$	0.8745 (20)	144	2.6		
p-Xylene	$C_6H_4(CH_3)_2$	0.8611 (20)	138.5	2.3		0.19 g/l. (25)

REFERENCES

1. ABBOT, C. D. and JOHNSON, E. I.: *Analyst* **82**, 206 (1957).
2. ADAM, J. A., BOOTH, E. and STRICKLAND, J. D. H.: *Anal. Chim. Acta* **6**, 462 (1952).
3. AHRLAND, S., CHATT, J. and DAVIES, N. R.: *Quart. Rev. (London)* **12**, 265 (1958).
4. AHRLAND, S., CHATT, J., DAVIES, N. R. and WILLIAMS, A. A.: *J. Chem. Soc.* **1958**, 264, 276, 1403.
5. ALIMARIN, I. P., GIBALO, I. M. and PIGAGA, A. K.: *Doklady Akad. Nauk SSSR* **186**, 1323 (1969).
6. ALIMARIN, I. P. and KRAUSZ, I.: *Magy. Kém. Foly.* **66**, 262 (1960).
7. ALIMARIN, I. P., PETRUKIN, O. M. and ZOLOTOV, YU. A.: *Zh. Anal. Khim.* **17**, 554 (1962).
8. ALIMARIN, I. P., PRZHEVALSKII, E. S., PUZDRENKOVA, I. V. and GOLOVINA, A. P.: *Trudy Komissii Anal. Khim. Akad. Nauk SSSR* **8**, 152 (1959).
9. ALIMARIN, I. P. and SHLENSKAYA, V. I.: *Pure Appl. Chem.* **21**, 461 (1970).
10. ALIMARIN, I. P., TIKHVINSKAYA, T. I., SHLENSKAYA, V. I. and BIRYUKOV, A. A.: *Izv. Akad. Nauk SSSR, Ser. Khim.* 2675 (1968).
11. ALMÁSY, A.: *Acta Chim. Acad. Sci. Hung.* **10**, 303 (1957); **17**, 55 (1958).
12. ALMÁSSY, GY. and VIGVÁRI, M.: *Magy. Kém. Foly.* **61**, 109 (1955).
13. ALTEN, F., WEILAND, H. and LOOFMAN, H.: *Angew. Chem.* **46**, 668 (1933).
14. ANDEREGG, G.: *Helv. Chim. Acta* **46**, 2397 (1963).
15. ASHBROOCK, A. W. and RITCEY, G. M.: *Canad. J. Chem.* **39**, 1109 (1961).
16. ATHAVALE, V. T., RAMACHANDRAN, T. P., TILLU, M. M. and VAIDYA, G. M.: *Anal. Chim. Acta* **24**, 263 (1961).
17. AYRES, G. H. and NARANG, B. D.: *Anal. Chim. Acta* **24**, 241 (1961).
18. AYRES, G. H. and YOUNG, F.: *Anal. Chem.* **22**, 1281 (1950).
19. BABKO, A. K.: *Zav. Lab.* **25**, 45 (1959).
20. BABKO, A. K. and PILIPENKO, A. T.: *Zh. Anal. Khim.* **2**, 33 (1947).
21. BABKO, A. K. and SHTOKALO, M. I.: *Ukr. Khim. Zh.* **30**, 220 (1964).
22. BAILEY, T. H. and LYLE, S. J.: *Talanta* **12**, 563 (1965).
23. BAMBACH, K.: *Ind. Eng. Chem. Anal. Ed.* **12**, 63 (1940).
24. BANKS, C. V.: *Analytical Chemistry*, Proc. Int. Symp. Birmingham, 1962, p. 131.
25. BANKS, C. V. and ANDERSON, S.: *J. Amer. Chem. Soc.* **84**, 1486 (1962).
26. BANKS, C. V. and ANDERSON, S.: *Inorg. Chem.* **2**, 112 (1963).
27. BANKS, C. V. and BARNUM, D. W.: *J. Amer. Chem. Soc.* **80**, 3579 (1958).
28. BÁNYAI, E.: *Kémiai indikátorok (Chemical Indicators)*. Műszaki Könyvkiadó (Technical Publisher), Budapest, 1961.
29. BARBER, H. H. and GRZESKOVIAK, E.: *Anal. Chem.* **21**, 192 (1949).
30. BARD, J. A.: *Chemical Equilibrium*, Harper–Row, 1966.
31. BARNES, H.: *Analyst* **76**, 220 (1951).
32. BARNES, J. W. and DOROUGH, G. D.: *J. Amer. Chem. Soc.* **72**, 4045 (1950).
33. BASOLO, F. and MURMANN, R. K.: *J. Amer. Chem. Soc.* **74**, 5243 (1952); **76**, 211 (1954).
34. BASSET, J., LETOW, T. B. and VOGEL, A. I.: *Analyst* **92**, 279 (1967).
35. BAYER, E.: *Angew. Chem.* **76**, 76 (1964).
36. BECK, G.: *Mikrochem.* **34**, 282 (1949).
37. BECK, M. T.: *Chemistry of Complex Equilibria*. Publishing House of the Hungarian Academy of Sciences and Van Nostrand Ltd., Budapest–London, 1970.
38. BECK, M. T. and GÖRÖG, S.: *Acta Chim. Acad. Sci. Hung.* **22**, 159 (1960).
39. BELCHER, R.: *Acta Chim. Acad. Sci. Hung.* **33**, 257 (1962).
40. BELCHER, R.: *Talanta* **12**, 129 (1965).

41. BELCHER, R.: *Proc. Soc. Analyt. Chem.* **5**, 4 (1968).
42. BELCHER, R.: *Talanta* **15**, 357 (1968).
43. BELCHER, R. and BETTERIDGE, D.: *Talanta* **13**, 535 (1966).
44. BELCHER, R., JENKINS, C. R., STEPHEN, W. I. and UDEN, P. C.: *Talanta* **17**, 455 (1970).
45. BELCHER, R., MAJER, J., PERRY, R. and STEPHEN, W. I.: *J. Inorg. Nucl. Chem.* **31**, 471 (1969).
46. BELCHER, R., STEPHEN, W. I. and UDEN, P. C.: *Anal. Chim. Acta* **39**, 357 (1967).
47. BERG, R.: *Z. anal. Chem.* **76**, 191 (1929).
48. BERG, R.: *Z. anorg. allgem. Chem.* **204**, 208, 215 (1932).
49. BERG, R.: *Die analytische Verwendung des o-Oxy-chinolins (Oxin) und seiner Derivate.* Stuttgart, 1938.
50. BERG, R. and ROEBLING, W.: *Angew. Chem.* **47**, 404 (1934); **48**, 597 (1935); *Ber.* **68**, 403 (1935).
51. BERGER, W. and ELVERS, H.: *Z. anal. Chem.* **171**, 185 (1959).
52. BERGER, W. and ELVERS, H.: *Z. anal. Chem.* **171**, 256 (1959).
53. BERSIN, TH.: *Z. anal. Chem.* **85**, 428 (1931).
54. BETTERIDGE, D.: *Talanta* **12**, 129 (1965).
55. BETTS, R. H. and LEIGH, R. M.: *Canad. J. Res.* **B28**, 514 (1950).
56. BEZZI, S., BUA, E. and SCHIAVINATO, G.: *Gazz. Chim. ital.* **81**, 856 (1951).
57. BILLO, E. J., ROBERTSON, B. E. and GRAHAM, R. P.: *Talanta* **10**, 757 (1963).
58. BIRYUKOV, A. A., SHLENSKAYA, V. I., ALIMARIN, I. P. and TIKHVINSKAYA, T. I.: *Zh. Neorg. Khim.* **9**, 1679 (1966).
59. BJERRUM, J.: *Metal Ammine Formation in Aqueous Solution.* Haase, Copenhagen, 1941.
60. BLAEDEL, W. J. and OLSEN, E. D.: *Anal. Chem.* **32**, 1867 (1960).
61. BLAKE, C. A., BAES, C. F., BROWN, K. B., COLEMAN, C. F. and WHITE, J. C.: *Proc. Int. Conf. Peaceful Uses of Atomic Energy, Geneva,* 1958, p. 289.
62. BLAU, F.: *Monatsh. Chem.* **19**, 647 (1898).
63. BLINC, R. and HADŽI, D.: *J. Chem. Soc.* **1958**, 4536.
64. BLINC, R. and HADŽI, D.: *Spectrochim. Acta* **16**, 852 (1960).
65. BOCHKOVA, V. M. and PESHKOVA, V. M.: *Zh. Neorg. Khim.* **3**, 1131 (1958).
66. BODE, H.: *Z. anal. Chem.* **142**, 414 (1954).
67. BODE, H.: *Z. anal. Chem.* **143**, 182 (1954).
68. BODE, H.: *Z. anal. Chem.* **144**, 90 (1955).
69. BODE, H.: *Z. anal. Chem.* **144**, 165 (1955).
70. BODE, H. and NEUMANN, F.: *Z. anal. Chem.* **172**, 1 (1960).
71. BODE, H. and TUSCHE, K. J.: *Z. anal. Chem.* **147**, 414 (1957).
72. BOLOMEY, R. A. and WISH, L.: *J. Amer. Chem. Soc.* **72**, 4483 (1950).
73. BORDNER, J., SALESIN, E. D. and GORDON, L.: *Talanta* **8**, 579 (1961).
74. BRAININA, K. Z., SAPOZHNIKOVA, E. YA. and KRAPIVKINA, T. A.: *Proc. IV. International Cong. Polarography, Prague, 1966.*
75. BRANDŠTETR, I. and VŘEŠTAL, J.: *Coll. Czech. Chem. Comm.* **26**, 392 (1961).
76. BRANNAN, J. R. and NANCOLLAS, G. H.: *Chem. Ind. (London)* **1959**, 1415.
77. BREANT, M.: *Bull. soc. chim. France,* 948 (1956).
78. BRISCOE, G. B. and COOKSEY, B. G.: *J. Chem. Soc. (A),* **1969**, 205.
79. BROWN, W. B., STEINBACH, J. F. and WAGNER, W. F.: *J. Inorg. Nucl. Chem.* **13**, 119 (1960).
80. BRYAN, R. F. and KNOPF, P. M.: *Proc. Chem. Soc.* **1961**, 203.
81. BUDESINSKY, B.: *Chelates in Analytical Chemistry.* M. Decker, New York (1969). Vol. II, p. 1
82. BULLWINKEL, E. P. and NOBLE, P.: *J. Amer. Chem. Soc.* **80**, 2955 (1958).
83. BURGER, K.: *Magy. Kém. Foly.* **67**, 208 (1961).
84. BURGER, K.: *Magy. Kém. Foly.* **67**, 519 (1961); *Talanta* **8**, 769 (1961).
85. BURGER, K.: *Talanta* **8**, 253 (1961); *Magy. Kém. Foly.* **67**, 349 (1961).
86. BURGER, K.: *Szerves fémreagensek alkalmazásának koordinációs kémiai és analitikai kérdései (Coordination Chemical and Analytical Problems of the Application of Organic Metal Reagents).* Publishing House of the Hungarian Academy of Sciences, Budapest, 1965. (In Hungarian.)
87. BURGER, K.: *Modern koordinációs kémiai vizsgáló módszerek (Modern Methods of Examination in Coordination Chemistry).* Publishing House of the Hungarian Academy of Sciences, Budapest, 1967. (In Hungarian.)

REFERENCES

88. BURGER, K.: Selectivity and analytical application of dimethylglyoxime. In: *Chelates in Analytical Chemistry*, Vol. II; M. Decker, New York, 1969.
88a. BURGER, K.: *Coordination Chemistry: Experimental Methods*. Butterworths, London 1973.
89. BURGER, K. and DYRSSEN, D.: *Acta Chem. Scand.* **17**, 1489 (1963).
90. BURGER, K. and EGYED, I.: *J. Inorg. Nucl. Chem.* **27**, 2361 (1965); *Magy. Kém. Foly.* **71**, 144 (1965)
91. BURGER, K., GAIZER, F. and SCHULEK, E.: *Talanta* **5**, 97 (1960).
92. BURGER, K., KORECZ, L., MANUABA, I. B. A. and MAG, P.: *J. Inorg. Nucl. Chem.* **28**, 1673 (1966); *Magy. Kém. Foly.* **72**, 224 (1966).
93. BURGER, K. and PINTÉR, B.: *J. Inorg. Nucl. Chem.* **29**, 1717 (1967).
94. BURGER, K. and RUFF, I.: *Talanta* **10**, 329 (1963); *Magy. Kém. Foly.* **69**, 240 (1963).
95. BURGER, K. and RUFF, I.: *Acta Chim. Acad. Sci. Hung.* **41**, 75 (1964); *Magy. Kém. Foly.* **70**, 226 (1964).
96. BURGER, K. and RUFF, I.: *Acta Chim. Acad. Sci. Hung.* **45**, 77 (1965).
97. BURGER, K., RUFF, I. and RUFF, F.: *Magy. Kém. Foly.* **70**, 394 (1964).
98. BURGER, K., RUFF, I. and RUFF, F.: *J. Inorg. Nucl. Chem.* **27**, 179 (1965); *Magy. Kém. Foly.* **70**, 351 (1964).
99. BURGER, K., RUFF, F., RUFF, I. and EGYED, I.: *Acta Chim. Acad. Sci. Hung.* **46**, 1 (1965).
100. BURGER, K., SYREK, G. and FARSANG, GY.: *Acta Chim. Acad. Sci. Hung.* **49**, 113 (1966); *Magy. Kém. Foly.* **72**, 380 (1966).
101. BURGER, K., SZÁNTÓ-HORVÁTH, G. and PAPP-MOLNÁR, E.: *Acta Chim. Acad. Sci. Hung.* **71**, 127 (1972)
102. BUSEV, A. I.: *Synthesis of New Organic Reagents for Inorganic Analysis*. Izd. MGU, 1970. (In Russian.)
103. BUSEV, A. I. and BAZHANOVA, L. A.: *Zh. Neorg. Khim.* **6**, 2210 (1961).
104. BUSEV, A. I., CHARLAMOV, I. P. and SMIRNOVA, O. W.: *Analyt. Letters* **3**, 177, 191 (1970).
105. BUSEV, A. I., GRÖSSEL, V. G. and IVANOV, V. M.: *Analyt. Letters* **1**, 267 (1968).
106. BUSEV, A. I., IVANOV, V. M. and BOGDANOVICH, L. I.: *Vestnik MGU*, Ser. II. Chem. **3**, 86 (1969).
107. BUSEV, A. I. and KISELEVA, L. V.: *Vestnik Moscow Univ., Ser. Mat., Mekh. Astr. Fiz. Khim.* **13**, 4, 179 (1958).
108. BUSEV, A. I. and POLIANSKII, N. G.: *The Use of Organic Reagents in Inorganic Analysis*. Pergamon Press, Oxford, 1960.
109. CAGLE, F. W. and SMITH, G. F.: *J. Amer. Chem. Soc.* **69**, 1860 (1947).
109a. CALVIN, M. and WILSON, K. W.: *J. Amer. Chem. Soc.* **67**, 2003 (1945).
110. CHAN, F. L.: *Talanta* **7**, 253 (1961).
111. CHANDELLE, R.: *Bull. soc. chim. Belg.* **50**, 185 (1941).
112. CHARLES, R. G. and FREISER, H.: *Anal. Chim. Acta* **11**, 101 (1954).
113. CHATT, J.: *J. Chem. Soc.* **1949**, 3340.
114. CHENG, K. L.: *Anal. Chem.* **30**, 1027 (1958).
115. CHENG, K. L.: *Anal. Chem.* **33**, 783 (1961).
116. CHENG, K. L.: *Talanta* **9**, 739 (1962).
117. CHENG, K. L. and BRAY, R. H.: *Anal. Chem.* **27**, 782 (1955).
118. CHENG, K. L., BRAY, R. H. and MELSTED, S. W.: *Anal. Chem.* **27**, 24 (1955).
119. CHERNIKOV, YU. A. and DOBKINA, B. M.: *Zav. Lab.* **25**, 131 (1959).
120. CHERNIKOV, YU. A., TRAMM, R. S. and PEVZNER, N. S.: *Zav. Lab.* **26**, 921 (1960).
121. CHERNITSKAYA, R. E.: *Zh. Obshch. Khim.* **22**, 408 (1952).
122. CHOPPIN, G. R. and SILVA, R. J.: *J. Inorg. Nucl. Chem.* **3**, 153 (1956).
123. CLARKE, K., COWEN, R. A., GRAY, G. W. and OSBORNE, E. H.: *J. Chem. Soc.* **1963**, 245.
124. CLAYTON, R. F., HARDWICK, W. H., MORETON-SMITH, M. and TODD, R.: *Analyst* **83**, 13 (1958).
125. CLIFFORD, P. A.: *J. Assoc. Offic. Agric. Chemists* **26**, 26 (1943).
126. CLUETT, M. L. and YOE, J. M.: *Anal. Chem.* **29**, 1265 (1957).
127. CONNICK, R. E. and MCVEY, W. H.: *J. Amer. Chem. Soc.* **71**, 3182 (1949).
128. COOK, E. H. and TAFT, R. V.: *J. Amer. Chem. Soc.* **74**, 6103 (1952).
129. COOPER, S. S. and SULLIVAN, M. L.: *Anal. Chem.* **21**, 1135 (1949).
130. CORKINS, J. T., PIETRZAK, R. F. and GORDON, L.: *Talanta* **9**, 49 (1962).

130a. Corsini, A., Fernando, Q. and Freiser, H.: *Anal. Chim.* **35**, 1424 (1963).
131. Cyrankowska, M.: *Chem. Anal. (Warsaw)* **6**, 649 (1961).
132. Davis, W. F.: *Talanta* **16**, 1330 (1969).
133. Dawson, E. C.: *Analyst* **73**, 618 (1948).
134. Day, A. R. and Taggart, W. T.: *Ind. Eng. Chem.* **20**, 545 (1928).
135. De, A. K. and Rahaman, M. S.: *Anal. Chem.* **27**, 591 (1962).
136. De Bruin, H. J. and Temple, R. B.: *Austral. J. Chem.* **15**, 153 (1962).
137. Dinstl, G. and Hecht, F.: *Mikrochim. Acta* **1962**, 321.
138. Duval, C.: *Inorganic Thermogravimetric Analysis*. Elsevier, New York, 1953.
139. Dyrssen, D.: *Svensk. Kem. Tidskr.* **65**, 43 (1953).
140. Dyrssen, D.: *Svensk. Kem. Tidskr.* **75**, 12 (1963).
141. Dyrssen, D. and Dahlberg, V.: *Acta Chem. Scand.* **7**, 1186 (1953).
142. Dyrssen, D. and Hennichs, M.: *Acta Chem. Scand.* **15**, 47 (1961).
143. Dyrssen, D., Jaguer, D. and Wengelin, F.: *Computer Calculations of Ionic Equilibria and Titration Procedures*. J. Wiley, London, 1968.
144. Dyrssen, D., Krasovec, F. and Sillén, L. G.: *Acta Chem. Scand.* **13**, 50 (1959).
145. Eckert, H. W.: *Ind. Eng. Chem. Anal. Ed.* **15**, 406 (1943).
146. Elliot, C. R., Preston, P. F. and Thompson, J. H.: *Analyst* **84**, 237 (1959).
147. Ephraim, F.: *Ber.* **63**, 1928 (1930).
148. Erdey, L.: *A kémiai analízis súlyszerinti módszerei (Gravimetric Methods of Chemical Analysis)*. Publishing House of the Hungarian Academy of Sciences, Budapest, 1960. (In Hungarian.)
149. Erdey, L., Buzás, I. and Vigh, K.: *Talanta* **1**, 377 (1958).
150. Erdey, L., Rády, G. and Fleps, V.: *Anal. Chem.* **135**, 1 (1962).
151. Erdeyné, Schneer, A.: *Magyar Kémikusok Lapja* **19**, 325 (1964).
152. Erdeyné, Schneer, A. and Tóth, T.: *Magy. Kém. Foly.* **71**, 244 (1965); *Annales Univ. Scient. R. Eötvös Bpest. Sectio Chim.* **7**, 11 (1965).
153. Eshelman, H. C., Dean, J. A., Menis, O. and Rains, T. S.: *Anal. Chem.* **31**, 183 (1959).
154. Fajans, K.: *Naturwiss.* **11**, 165 (1922).
155. Feigl, F.: *Mikrochemie* **1**, 74 (1923); *Z. anal. Chem.* **64**, 41 (1924).
156. Feigl, F.: *Ber.* **57**, 758 (1924).
157. Feigl, F.: *Z. anal. Chem.* **74**, 380 (1928).
158. Feigl, F.: *Chemistry of Specific, Selective and Sensitive Reactions*. Academic Press, New York, 1949.
159. Feigl, F. and Miranda, L. I.: *Ind. Eng. Chem. Anal. Ed.* **16**, 141 (1944).
160. Feigl, F. and Ordelt, H.: *Z. anal. Chem.* **65**, 448 (1925).
161. Feigl, F., Sicher, G. and Singer, O.: *Ber.* **58**, 2294 (1925).
161a. Fernando, Q. and Freiser, H.: *J. Amer. Chem. Soc.* **80**, 4928 (1958).
162. Firsching, F. H. and Brewer, J. G.: *Anal. Chem.* **35**, 1630 (1963).
163. Fischer, H.: *Angew. Chem.* **47**, 685 (1934).
164. Fischer, H.: *Angew. Chem.* **50**, 919 (1937).
165. Fischer, H. and Leopoldi, G.: *Mikrochim. Acta* **1**, 30 (1937).
166. Fischer, H., Leopoldi, G. and Von Uslar, H.: *Z. anal. Chem.* **101**, 1 (1935).
167. Flagg, J. F. and Furman, N. H.: *Ind. Eng. Chem. Anal. Ed.* **12**, 529, 663 (1940).
168. Flaschka, H.: *Chemist Analyst* **44**, 2 (1955).
169. Fleck, G. M.: *Equilibria in Solution*. Holt, Rinehart Winston, 1966.
170. Forster, W. A.: *Analyst* **78**, 613 (1953).
171. Fortune, W. B. and Mellon, M. G.: *Ind. Eng. Chem. Anal. Ed.* **10**, 60 (1938).
172. Frasson, E., Bardi, R. and Bezzi, S.: *Acta Cryst.* **12**, 201 (1959).
173. Frasson, E., Bardi, R., Zanetti, R. and Mammi, M.: *Ann. di Chim.* **48**, 1027 (1958).
174. Frasson, E., Panattoni, C. and Zanetti, R.: *Acta Cryst.* **12**, 1027 (1959).
175. Freiser, H.: *Analyst* **74**, 830 (1952).
176. Freiser, H. and Fernando, Q.: *Ionic Equilibria in Analytical Chemistry*. John Wiley, New York, 1963.
177. Freiser, B. S. and Freiser, H.: *Talanta* **17**, 540 (1970).
178. Freiser, B. S. and Freiser, H.: *Anal. Chem.* **42**, 305 (1970).
179. Freiser, H. and Morrison, G. H.: *Ann. Rev. Nucl. Sci.* **9**, 227 (1959).
180. Friedeberg, H.: *Anal. Chem.* **27**, 305 (1955).

181. FUJITA, J., NAKAHARA, A. and TSUCHIDA, D.: *J. Chem. Phys.* **23**, 1541 (1955).
182. FUNK, H. and DITT, M.: *Z. anal. Chem.* **91**, 332, 241 (1933).
183. FUNK, H. and DITT, M.: *Z. anal. Chem.* **93**, 241 (1933).
184. FURMANN, N. H. and FLAGG, J. F.: *Ind. Eng. Chem. Anal. Ed.* **12**, 738 (1940).
185. GAHLER, A. R.: *Anal. Chem.* **26**, 577 (1954).
186. GALLEGO, R., DEIJS, W. B. and FELDMEIJER, J. H.: *Rec. trav. chim.* **71**, 987 (1952).
187. GARDNER, K.: *Analyst* **76**, 485 (1951).
188. GENTRY, C. H. R. and SHERRINGTON, L. G.: *Analyst* **71**, 432 (1946).
189. GENTRY, C. H. R. and SHERRINGTON, L. G.: *Analyst* **75**, 17 (1950).
190. GERBER, L., CLAASSEN, R. I. and BORUFF, C. S.: *Ind. Eng. Chem. Anal. Ed.* **14**, 658 (1942).
191. GIBALO, I. M., ALIMARIN, I. P. and DAVAADORZH, P.: *Zh. Anal. Khim.* **19**, 835 (1963).
192. GIBALO, I. M., ALIMARIN, I. P. and EREMINA, G. V.: *Doklady Akad. Nauk SSSR* **176**, 1300 (1967).
193. GILLARD, R. D. and WILKINSON, G.: *J. Chem. Soc.* **1963**, 6041.
194. GODYCKI, L. E. and RUNDLE, R. E.: *Acta Cryst.* **6**, 487 (1959).
195. GOLDSTEIN, G., MANNING, D. L. and MENIS, O.: *Anal. Chem.* **31**, 192 (1959).
196. GORJUSINA, V. G., ROMANOVA, E. V. and ARCHAKOVA, T. A.: *Zav. Lab.* **27**, 795 (1961).
197. GYENES, I.: *Titrations in Non-Aqueous Media*. Publishing House of the Hungarian Academy of Sciences, Budapest, 1967.
198. HADDOCK, L. A.: *Analyst* **59**, 163 (1934).
199. HAGEMANN, F.: *J. Amer. Chem. Soc.* **72**, 768 (1950).
200. HAHN, R. B.: *Anal. Chem.* **21**, 1579 (1949).
201. HAHN, R. B. and BAGINSKI, E. S.: *Anal. Chim. Acta* **14**, 45 (1956).
202. HALL, A. J. and YOUNG, R. S.: *Analyst* **71**, 479 (1946).
203. HAMMES, G. H. and STENFELD, V. I.: *J. Amer. Chem. Soc.* **84**, 4639 (1969).
204. HARGIS, L. G. and BOLTZ, T. F.: *Talanta* **11**, 57 (1964).
205. HARTMANN, H.: *Theorie der chemischen Bindung und Quantum-theoretische Grundlagen*. Springer Verlag, 1956.
206. HASHITANI, H. and YAMAMOTO, K.: *J. Chem. Soc. Japan (Pur. Chem. Sect.)* **80**, 727 (1959).
207. HENRIQUES, R. and ILINSKI, J.: *Ber.* **17**, 2592 (1884); **18**, 699 (1885).
208. HILEMAN, O. E., ELLEFSEN, P. R., MAGEE, R. J. and GORDON, L.: *Talanta* **10**, 419 (1963).
209. HINSWARK, O. N., HOUFF, W. H., WITTWER, S. H. and SELL, H. M.: *Anal. Chem.* **26**, 1202 (1954).
210. HOLZER, H.: *Z. anal. Chem.* **95**, 392 (1933).
211. HUBBARD, D. M.: *Anal. Chem.* **20**, 363 (1948).
212. HUFFMAN, E. H. and BEAUFAIT, L. J.: *J. Amer. Chem. Soc.* **71**, 3179 (1949).
213. HULANICKI, A.: *Talanta* **14**, 1371 (1967).
214. HYNEK, R. J. and WRANGELL, L. J.: *Anal. Chem.* **28**, 1520 (1956).
215. INCZÉDY, J.: *Ioncserélők analitikai alkalmazása (Analytical Applications of Ion-exchangers)*. Műszaki Könyvkiadó (Technical Publisher), Budapest, 1962 (In Hungarian.)
216. INCZÉDY, J.: *Analytical Applications of Ion Exchangers*. Pergamon Press, Oxford, 1966.
217. INCZÉDY, J.: *Komplex egyensúlyok analitikai alkalmazása (Analytical Application of Complex Equilibria)*. Műszaki Könyvkiadó (Technical Publisher), Budapest, 1970 (In Hungarian.)
218. IRVING, H.: *Quart. Rev.* **5**, 200 (1951).
219. IRVING, H. and BELL, C. F.: *J. Chem. Soc.* **1952**, 1216.
220. IRVING, H. and BELL, C. F.: *J. Chem. Soc.* **1953**, 3538.
221. IRVING, H., BELL, C. F. and WILLIAMS, R. J. P.: *J. Chem. Soc.* **1952**, 356
222. IRVING, H. and MELLOR, D. H.: *J. Chem. Soc.* **1962**, 5222.
223. IRVING, H. and PETTIT, L.: *Analytical Chemistry*. Proc. Int. Symp. Birmingham, 1962, p. 122.
224. IRVING, H. and RAMAKRISHNA, R. S.: *Analyst* **85**, 860 (1960)
225. IRVING, H. and ROSOTTI, H. S.: *J. Chem. Soc.* **1954**, 2910.
226. IRVING, H. and ROSOTTI, H. S.: *Analyst* **80**, 245 (1955).
227. IRVING, H. and WILLIAMS, R. J. P.: *Nature* **162**, 746 (1948).
228. IRVING, H. and WILLIAMS, R. J. P.: *Analyst* **77**, 813 (1952).
229. ISHIHARA, Y., SHIBATA, K., KISHI, H. and HORI, T.: *Japan Analyst (Bunseki Kagaku)* **11**, 91 (1962).

230. IWANTSCHEFF, G.: *Das Dithizon und seine Anwendung in der Mikro- und Spurenanalyse.* Weinheim, 1958.
231. JASIM, F., MAGEE, R. J. and WILSON, C. L.: *Rec. trav. chim.* **79**, 541 (1960).
232. JENNINGS, J.: *J. Chem. Soc.* **1935**, 818.
233. JILLOT, B. A. and WILLIAMS, R. J. P.: *J. Chem. Soc.* **1958**, 462.
234. JONES, J. P., HILEMAN, O. E. and GORDON, L.: *Talanta* **10**, 111 (1963).
235. JONES, J. P., HILEMAN, O. E., TOWNSEND, A. and GORDON, L.: *Talanta* **11**, 855 (1964).
236. JONES, J. L. and HOWICK, L. C.: *Talanta* **11**, 757 (1964).
237. JONES, J. G., POOL, J. B., TOMKINSON, J. C. and WILLIAMS, R. J. P.: *J. Chem. Soc.* **1958**, 2001.
238. JØRGENSEN, C. K.: *Energy Levels of Complexes and Gaseous Ions.* Gjellerups, Copenhagen, 1957.
239. KALINKIN, P. and SEMIKOZOV, G. S.: *Zav. Lab.* **27**, 17 (1961).
240. KANNER, L. J., SALESIN, E. D. and GORDON, L.: *Talanta* **7**, 288 (1963).
241. KAWAHATA, M., MOCHIZUKI, M. and MISAKI, T.: *Japan Analyst (Bunseki Kagaku)* **11**, 1017 (1962).
242. KAWAHATA, M., MOCHIZUKI, M. and MISAKI, T.: *Japan Analyst (Bunseki Kagaku)* **11**, 1020 (1962).
243. KEENAN, T. K. and SUTTLE, J. F.: *J. Amer. Chem. Soc.* **76**, 2184 (1954).
244. KHOPKAR, S. M.: *Z. anal. Chem.* **171**, 241 (1959).
245. KHOPKAR, S. M. and DE, A. K.: *Chem. Ind.* **1959**, 291.
246. KIMURA, K. and MURAKAMI, Y.: *Mikrochem.* **36/37** (1951).
247. KIMURA, K., SAITO, K. and ASADRA, M.: *Bull. Chem. Soc. Japan* **29**, 640 (1956).
248. KIRSCHNER, S. and BHATNAGER, D. C.: *Anal. Chem.* **35**, 1069 (1963).
249. KISS, A. and ALMÁSSY, GY.: *Magy. Kém. Foly.* **64**, 332 (1958).
250. KLEMENT, R.: *Z. anal. Chem.* **128**, 431 (1948).
251. KNOWLES, H.: *Bur. Stand. J. Research* **9**, 1 (1932).
252. KOLARIK, Z. and PÁNKOVA, H.: *Coll. Czech. Chem. Comm.* **27**, 166 (1962).
253. KOROLEFF, F.: *Merentutkimuslaitoksen Julkaisu Havsforsking-institutes Skrift (Helsinki)* **145**, 7 (1950).
254. KRAUSZ, I.: *Magy. Kém. Foly.* **66**, 218, 296, 344 (1960).
255. KREIMER, S. E. and BUTYLKIN, L. P.: *Zav. Lab.* **24**, 131 (1958).
256. KRISHEN, A. and FREISER, H.: *Anal. Chem.* **29**, 288 (1957).
257. KRISHEN, A. and FREISER, H.: *Anal. Chem.* **31**, 923 (1959).
258. KULJBERG, L. M.: *Organicheski Reaktivi v Analiticheskoi khimii.* Goshhimizdat, 1950.
259. KUMINS, CH. A.: *Anal. Chem.* **19**, 376 (1947).
260. KUMINS, CH. A.: *Anal. Chem.* **21**, 1352 (1949).
261. KUZNETSOV, V. I.: *Zh. Anal. Khim.* **14**, 7 (1959).
262. KUZNETSOV, V. I., KORENMAN, I. M. and KULJBERG, L. M.: *Trudi Kom. Anal. Khim.* **5**, (8) 3 (1954).
263. KUZNETSOV, V. I., MYASOEDOVA, C. M., NI, G. V. and OKHANOVA, L. A.: *Hua Hsueh Hsueh Pao*, **27**, 74 (1961).
264. LACOSTE, R. J., EARING, M. H. and WIBERLY, S. E.: *Anal. Chem.* **23**, 871 (1951).
265. LAI, M. G. and WEISS, H. V.: *Anal. Chem.* **34**, 1012 (1962).
266. LAING, M. and ASOP, P. A.: *Talanta* **17**, 242 (1970).
267. LERNER, M. and RIEMAN, W.: *Anal. Chim. Acta* **13**, 439 (1955).
268. LEWIS, J. and WILKINS, R. G.: *Modern Coordination Chemistry.* Interscience Publ., New York, 1960.
269. LINDNER, M.: U.S. Atomic Energy Commission Report, UCRL 4377.
270. LIPPARD, S. J.: *J. Amer. Chem. Soc.* **88**, 4300 (1966).
271. LIPPARD, S. J., COTTON, F. A. and LEGZDINS, P.: *J. Amer. Chem. Soc.* **89**, 5930 (1967).
272. LUKE, C. L.: *Anal. Chem.* **28**, 1443 (1956).
273. LUKE, C. L. and CAMPBELL, M. E.: *Anal. Chem.* **26**, 1778 (1954).
274. LUKYANOV, V. F., SAVVIN, S. B. and NIKOLSKAYA, I. B.: *Zh. Anal. Khim.* **15**, 311 (1960).
275. LUMME, P. O.: *Suomen Kem.* **33B**, 69, 85 (1960).
276. LYLE, S. J. and SOUTHERN, D. L.: *Talanta* **11**, 1239 (1964).
277. MAGNUSSON, L. B. and ANDERSON, M. L.: *J. Amer. Chem. Soc.* **76**, 6207 (1954).
278. MAIRANOVSKII, S. G.: *Electroanal. Chem.* **6**, 77 (1963).

REFERENCES

279. MAJUMDAR, A. K.: *Anal. Chem.* **32**, 1337 (1960).
280. MALISSA, H.: *Anal. Chim. Acta* **27**, 402 (1962).
281. MARCUS, Y.: *Acta Chem. Scand.* **11**, 329 (1957).
282. MAREC, D. J., SALESIN, E. D. and GORDON, L.: *Talanta* **8**, 293 (1961).
283. MARGERUM, D. W., BYRD, C. H., REED, S. A. and BANKS, C. V.: *Anal. Chem.* **25**, 1219 (1953).
284. MARGERUM, D. W., ZABIN, B. A. and JONES, D. L.: *Inorg. Chem.* **5**, 250 (1966).
285. MARSTON, H. R. and DEWEY, D. W.: *Australian J. Exp. Biol. Med. Sci.* **18**, 343 (1940).
286. MARTELL, A. E. and CALVIN, M.: *Chemistry of the Metal Chelate Compounds*. Prentice Hall, New York, 1953.
287. MATH, K. S., BHATKI, K. S. and FREISER, H.: *Talanta* **16**, 412 (1969).
287a. MATH, K. S., FERNANDO, Q. and FREISER, H.: *Anal. Chem.* **36**, 1762 (1964).
288. MATH, K. S. and FREISER, H.: *Anal. Chem.* **41**, 1682 (1969).
289. MAY, I. and HOFFMANN, J. I.: *J. Washington Acad. Sci.* **38**, 329 (1948).
290. MAYR, C.: *Z. anal. Chem.* **98**, 402 (1934).
291. MAYR, C. and FEIGL, F.: *Z. anal. Chem.* **90**, 15 (1932).
292. MCKAVENEY, J. P. and FREISER, H.: *Anal. Chem.* **29**, 290 (1957).
293. MCKAVENEY, J. P. and FREISER, H.: *Anal. Chem.* **30**, 526 (1958).
294. MEITES, L.: *Polarographic Techniques*. Interscience Publ., New York, 1955.
295. MERRITT, L. L. and WALKER, J. K.: *Ind. Eng. Chem. Anal. Ed.* **16**, 387 (1944).
296. MEYER, K. H. and HOPFF, H.: *Ber.* **54**, 579 (1921).
297. MEYER, S. and KOCH, O. G.: *Mikrochim. Acta* **1958**, 744.
298. MILLER, F. F., GEDDA, K. and MALISSA, H.: *Mikrochem.* **40**, 373 (1953).
299. MILLER, W. L. and WACHTER, L. E.: *Anal. Chem.* **22**, 1312 (1950).
300. MIYAMOTO, M.: *Japan Analyst (Bunseki Kagaku)* **9**, 925 (1960).
301. MOELLER, T.: *Ind. Eng. Chem. Anal. Ed.* **15**, 270 (1943).
302. MOELLER, T.: *Ind. Eng. Chem. Anal. Ed.* **15**, 346 (1943).
303. MOELLER, T. and COHEN, A. J.: *Anal. Chem.* **22**, 686 (1950).
304. MOORE, F. L.: *Anal. Chem.* **28**, 997 (1956).
305. MOORE, F. L.: *Anal. Chem.* **29**, 941 (1957).
306. MOORE, F. L., FAIRMAN, W. O., GANCHOF, J. G. and SURAK, J. G.: *Anal. Chem.* **31**, 1148 (1959).
307. MOORE, F. L. and HUDGENS, J. E.: *Anal. Chem.* **29**, 1767 (1957).
308. MORET, R. and BRUNISHOLZ, G.: *Z. anal. Chem.* **187**, 137 (1962).
309. MORGAN, G. and BURSTALL, F.: *J. Chem. Soc.* **135**, 20 (1932).
310. MORIMOTO, Y., ASHIZAWA, T. and ARAYA, S.: *Japan Analyst (Bunseki Kagaku)* **10**, 1387 (1961).
311. MORIMOTO, Y., ASHIZAWA, T. and MIYAHARA, K.: *Japan Analyst (Bunseki Kagaku)* **11**, 61 (1962).
312. MORRISON, G. H. and FREISER, H.: *Solvent Extraction in Analytical Chemistry*. John Wiley, New York, 1957.
313. MOSHIER, R. W. and SIEVERS, R. E.: *Gas Chromatography of Metal Chelates*. Pergamon Press, Oxford, 1965.
314. MOSS, M. L. and MELLON, M. G.: *Ind. Eng. Chem. Anal. Ed.* **14**, 862 (1942).
315. MOTOJIMA, K., HASHITANI, H. and YOSHIDA, H.: *Japan Analyst (Bunseki Kagaku)* **10**, 79 (1961).
316. MOTOJIMA, K., HASHITANI, H. and YOSHIDA, H.: *Japan Analyst (Bunseki Kagaku)* **11**, 659 (1962).
317. MUCHINA, Z. S., TIKHONOVA, A. A. and ZHEMCHUZHNAYA, I. A.: *Trudy Kom. Analit. Khim.* **12**, 298 (1960).
318. MYASSOEDOV, B. and MUXARD, R.: *Bull. soc. chim. France* **1962**, 237.
319. NAKAHARA, A.: *Bull. Chem. Soc. Japan* **28**, 473 (1955).
320. NAKAHARA, A., FUJITA, J. and TSUCHIDA, R.: *Bull. Chem. Soc. Japan* **29**, 296 (1956).
321. NANCE, K. W.: *Anal. Chem.* **23**, 1034 (1951).
322. NÄSÄNEN, R.: *Suomen Kem.* **26B** 2, 11 (1953).
323. NAZARENKO, V. A. and SHITAREVA, G. G.: *Zav. Lab.* **25**, 28 (1959).
324. NEDELJAK, M.: *Picrolonic Acid*. Chemapol, Prague, 1959.
325. NIELSCH, W.: *Z. anal. Chem.* **150**, 114 (1956).

326. NORWITZ, G., COHEN, J. and EVERETT, M. E.: *Anal. Chem.* **32**, 1132 (1960).
327. NOVÁK, J. V. A., KUTA, J. and RIHA, J.: *Chem. Listy* **47**, 649 (1953).
328. OKAČ, A.: *Coll. Czech. Chem. Comm.* **10**, 177 (1938).
329. *Organicheskie Reagenti v Analiticheskoi Khimi.* Izd. Nauk SSSR, Moscow, 1960.
330. ORGEL, L. E.: *An Introduction to Transition Metal Chemistry.* John Wiley, New York, 1960
331. PALEJ, P. N.: *Zh. Anal. Khim.* **15**, 598 (1960).
332. PALMA, R. J., REINBOLD, P. E. and PEARSON, K. H.: *Anal. Chem.* **42**, 47 (1970).
333. PANATTONI, C., FRASSON, E. and ZANETTI, R.: *Gazz. Chim. Ital.* **89**, 2132 (1959).
334. PAULING, L.: *The Nature of the Chemical Bond.* Ithaca, New York, 1948.
335. PAULING, L.: *J. Chem. Soc.* **1948**, 1461.
336. PEARSON, R. G.: *J. Amer. Chem. Soc.* **85**, 3533 (1963).
337. PEARSON, R. G.: *Chemistry in Britain* **3**, 103 (1967).
338. PERKINS, M. and REYNOLDS, G. F.: *Anal. Chim. Acta* **18**, 616, 625 (1958); **19**, 54, 194 (1959).
339. PERRIN, D. D.: *Organic Complexing Reagents.* Interscience Publ., New York, 1964.
340. PESHKOVA, V. M., BOCHKOVA, V. M. and ASTAHOVA, E. K.: *Zh. Anal. Khim.* **16**, 596 (1961).
341. PESHKOVA, V. M. and IGNATEVA, N. G.: *Zh. Anal. Khim.* **17**, 1087 (1962).
342. PESHKOVA, V. M., SHLENSKAYA, V. I. and SOKOLOV, S. S.: *Trudy. Kom. Anal. Khim.* **11**, 328 (1960).
343. PESHKOVA, V. M. and ZOZULYA, A. P.: *Nauchn. Dokl. Vis. Shkoly Khim. i Khim. Tekhnol.* **1**, 470 (1958).
344. PETERSON, S.: *J. Inorg. Nucl. Chem.* **14**, 126 (1960).
345. *Pharmacopoea Hung. V*, Vol. I, p. 108.
346. PILIPENKO, A. T.: *Zh. Anal. Khim.* **5**, 14 (1950).
347. PILIPENKO, A. T.: *Zh. Anal. Khim.* **8**, 286 (1953).
348. PILIPENKO, A. T. and YEREMENKO, O. M.: *Ukr. Khim. Zh.* **29**, 532 (1963).
349. PIRTEA, T. I.: *Rev. Chim. (Bucuresti)* **12**, 223 (1961); *Chem. Abstr.* **57**, 34a (1962)
350. PIRTEA, T. I.: *Rev. Chim. (Bucuresti)* **13**, 234 (1962); *Chem. Abstr.* **57**, 15792d (1962).
351. PIRTEA, T. I.: *Zh. Analit. Khim.* **20**, 824 (1965); *Chem. Abstr.*, **63**, 17136d (1965).
352. PIRTEA, T. I., CRIVAT, D. and BLANARU, E.: *Analele Univ. "C. I. Parhon", Ser. Stiint. Nat.* **11**, 96 (1962); *Chem. Abstr.* **60**, 13867a (1964).
353. PIRTEA, T. I. and DUMITRU, M.: *Analele Univ. "C. I. Parhon", Ser. Stiint. Nat.* **10**, 211 (1961); *Chem. Abstr.* **58**, 7357b (1963).
354. PIRTEA, T. I. and SIMION-NICOLAU, J.: *Analele Univ. "C. I. Parhon", Ser. Stiint. Nat.* **11**, 101 (1962); *Chem. Abstr.* **60**, 13869a (1964).
355. PODTZAYNOVA, V. H.: *Trudy Kom. Anal. Khim.* **11**, 146 (1960).
356. POLUEKTOFF, N. S.: *Mikrochemie* **19**, 265 (1934).
357. POSKANZER, A. M. and FOREMAN, B. M.: *J. Inorg. Nucl. Chem.* **16**, 323 (1961).
358. PŘIBIL, R.: *Kompleksoni v khimicheskom Analize*, Inost. Lit., Moscow, 1960.
359. PŘIBIL, R.: *Komplexone in der chemischen Analyse*, VEB Deutscher Verlag der Wiss., Berlin, 1961.
360. PŘIBIL, R. and BURGER, K.: *Talanta* **4**, 8 (1960).
361. PŘIBIL, R. and JENIK, J.: *Coll. Czech. Chem. Comm.* **19**, 470 (1954).
362. PŘIBIL, R., JENIK, J. and KOBROVÁ, M.: *Chem. Listy* **46**, 603 (1952).
363. PŘIBIL, R., KOBROVÁ, M. and JENIK, J.: *Chem. Listy* **47**, 842 (1953).
364. *Proc. Symp. Coord. Chem., Amsterdam 1955*, p. 283.
365. PROSZT, J. and PAULIK, J.: *Magy. Kém. Foly.* **58**, 113 (1952).
366. PRUE, J. and SCHWARZENBACH, G.: *Helv. Chim. Acta* **33**, 963 (1950).
367. RABINOWITCH, E. and STOCKMAYER, W. H.: *J. Amer. Chem. Soc.* **64**, 335 (1942).
368. RAFF, P. and BROTZ, W.: *Z. anal. Chem.* **133**, 241 (1951).
369. RÄKER, K. O.: *Z. anal. Chem.* **173**, 57 (1960).
370. RAO, B. S. K., SARMA, D. V. N. and RAO, B. S. V. R.: *Z. anal. Chem.* **160**, 351 (1958).
371. RAY, P.: *Z. anal. Chem.* **79**, 94 (1929).
372. RAY, P. and BOSE, M. K.: *Z. anal. Chem.* **95**, 400 (1933).
373. RAY, P. and GUPTA, J.: *Mikrochemie* **17**, 14 (1935).
374. REID, J. C. and CALVIN, M.: *J. Amer. Chem. Soc.* **72**, 2948 (1950).
375. REINBOLD, P. E. and PEARSON, K. H.: *Talanta* **17**, 391 (1970).
376. RICE, A. C., FOGG, H. C. and JAMES, C.: *J. Amer. Chem. Soc.* **48**, 895 (1926).
377. RINGBOM, A.: *Finska Kemintsamfundets Medd.* **65**, 82 (1956).

378. Ringbom, A.: *J. Chem. Ed.* **35**, 282 (1958).
379. Ringbom, A.: *Complexation in Analytical Chemistry.* Interscience Publ., New York, 1963.
380. Ringbom, A. and Wänninen, E.: *Anal. Chim. Acta* **11**, 153 (1954).
381. Rollet, P.: *Compt. rend.* **183**, 212 (1926).
382. Rooney, R. C.: *Analyst* **83**, 546 (1958).
383. Rudenko, N. P. and Starý, J.: *Trudy Kom. Anal. Khim.* **9**, 28 (1958).
384. Rüdorf, W. and Zannier, H.: *Z. anal. Chem.* **137**, 1 (1952).
385. Rydberg, J.: *Svensk. Kem. Tidskr.* **62**, 179 (1950).
386. Rydberg, J.: *Arkiv Kemi.* **9**, 95 (1956).
387. Rydberg, J. and Rydberg, B.: *Arkiv Kemi.* **9**, 81 (1956).
388. Sachdev, S. L. and Wert, P. W.: *Atmos. Environ.* **2**, 331 (1968).
389. Sajó, I.: *Komplexometria (Complexometry).* Műszaki Könyvkiadó (Technical Publisher), Budapest, 1959. (In Hungarian.)
390. Salesin, E. D., Abrahamson, E. W. and Gordon, L.: *Talanta* **9**, 699 (1962).
391. Salesin, E. D. and Gordon, L.: *Talanta* **4**, 75 (1960).
392. Samuelson, O., Lunden, L. and Schramm, K.: *Z. anal. Chem.* **140**, 330 (1953).
393. Samuelson, O. and Sjöström, E.: *Anal. Chem.* **26**, 1908 (1954).
394. Samuelson, O., Sjöström, E. and Forblom, S.: *Z. anal. Chem.* **144**, 323 (1955).
395. Sandell, E. B.: *Ind. Eng. Chem. Anal. Ed.* **8**, 336 (1936).
396. Sandell, E. B.: *Ind. Eng. Chem. Anal. Ed.* **13**, 844 (1941).
397. Sandell, E. B.: *J. Amer. Chem. Soc.* **72**, 4660 (1950).
398. Sandell, E. B.: *Colorimetric Determination of Traces of Metals.* Interscience Publ., New York, 1950.
399. Sandell, E. B. and Perlich, R. W.: *Ind. Eng. Chem. Anal. Ed.* **11**, 309 (1939).
400. Savvin, S. B.: *Dokl. Akad. Nauk SSSR* **127**, 1231 (1959).
401. Savvin, S. B.: *Usp. Khim.* **32**, 195 (1963).
402. Savvin, S. B.: *Arzenazo III.* Atomizdat, Moscow, 1966.
403. Savvin, S. B. and Bagrejev, V. V.: *Zav. Lab.* **26**, 412 (1960).
404. Savvin, S. B. and Muk, A. A.: *Bull. Inst. Nucl. Sci.,* "*Boris Kidrich*" *(Beograd)* **12**, 97 (1961).
405. Sawada, T. and Kato, S.: *Japan Analyst (Bunseki Kagaku)* **11**, 544 (1962).
406. Schneer-Erdey, A.: *Talanta* **10**, 591 (1963).
407. Schneer, A. and Halmos, T.: *Magy. Kém. Foly.* **64**, 371 (1958).
408. Schneer, A. and Hartmann, H.: *Magy. Kém. Foly.* **65**, 31 (1959); *Acta Chim. Acad. Sci. Hung.* **22**, 35 (1960).
409. Schneer, A. and Hartmann, H.: *Magy. Kém. Foly.* **65**, 64 (1959); *Acta Chim. Acad. Sci. Hung.* **22**, 139 (1960).
410. Schneer, A. and Hartmann, H.: *Magy. Kém. Foly.* **67**, 309 (1961).
411. Schneer, A. and Patoh, P.: *Magy. Kém. Foly.* **67**, 334 (1961).
412. Schulek, E. and Laszlovszky, J.: *Mikrochim. Acta* **41** (1961).
413. Schwarzenbach, G.: *Helv. Chim. Acta* **33**, 974 (1950).
414. Schwarzenbach, G.: *Helv. Chim. Acta* **35**, 2344 (1952).
415. Schwarzenbach, G.: *Die komplexometrische Titration,* II. Ed. F. Enke, Stuttgart, 1956, p. 17.
416. Schwarzenbach, G., Ackermann, H., Maissen, B. and Anderegg, G.: *Helv. Chim. Acta* **35**, 2337 (1952).
417. Schwarzenbach, G. and Freitag, E.: *Helv. Chim. Acta* **34**, 1492 (1951).
418. Schwarzenbach, G. and Gut, R.: *Helv. Chim. Acta* **39**, 1589 (1956).
419. Schwarzenbach, G., Gut, R. and Anderegg, G.: *Helv. Chim. Acta* **37**, 936 (1954).
420. Schwarzenbach, G., Maissen, B. and Ackermann, H.: *Helv. Chim. Acta* **35**, 2333 (1952).
421. Schwarzenbach, G. and Moser, P.: *Helv. Chim. Acta* **36**, 581 (1953).
422. Schweitzer, G. K. and Dawidson, J. E.: *Anal. Chim. Acta* **35**, 467 (1966).
423. Schweitzer, G. K. and Randolph, D. R.: *Anal. Chim. Acta* **26**, 567 (1962).
423a. Sekido, E., Fernando, Q. and Freiser, H.: *Anal. Chem.* **36**, 1762 (1964).
424. Seyanova, F. R. and Kozhokina, G. J. A.: *Trudy Khim. i Khim. Tekhnol.* **3**, 70 (1960).
425. Sheperd, E. and Meinke, W. W.: U.S. Atomic Energy Commission Report AECR 3879.
426. Shibata, S.: *Anal. Chim. Acta* **22**, 479 (1960); **23**, 367, 434 (1960).
427. Shibata, S.: *Anal. Chim. Acta* **25**, 348 (1961).

428. SHIGEMATSU, T. and TABUSHI, M.: *Bull. Inst. Chem. Res. Kyoto Univ.* **39**, 35 (1961).
429. SHLENSKAYA, V. I., TIKHVINSKAYA, T. I., BIRYUKOV, A. A. and ALIMARIN, I. P.: *Izv. Akad. Nauk SSSR, Ser. Khim.* 2141 (1967).
430. SIDGWICK, N. V.: *J. Chem. Soc.* **1941**, 433.
431. SIEVERS, R. E., PONDER, B. W., MORRIS, M. L. and MOSHIER, R. W.: *Inorg. Chem.* **2**, 693 (1963).
432. SILL, C. W. and PETERSON, H. E.: *Anal. Chem.* **21**, 1268 (1949).
433. SILLÉN, L. G. and MARTELL, A. E.: *Stability Constants of Metal-Ion Complexes.* The Chemical Society, London 1964.
434. SIN'YAKOVA, S. I. and TVESTKOVA, L. A.: *Trudy Kom. Anal. Khim.* **12**, 191 (1960).
435. SMITH, G. F.: *Cupferron and Neo-cupferron.* G. F. Smith Co., Columbus, 1938.
436. SNYDER, L. J.: *Anal. Chem.* **19**, 684 (1947).
437. SONE, K. and KATO, M.: *Naturwiss.* **45**, 10 (1958).
438. SPIKE, C. G. and PARRY, R. W.: *J. Amer. Chem. Soc.* **75**, 2726, 3770 (1953).
439. SPITZER, M. and MEITES, L.: *Anal. Chim. Acta* **26**, 58 (1962).
440. SPRINGER, C. S., MEEK, D. W. and SIEVERS, R. E.: *Inorg. Chem.* **6**, 1105 (1967).
441. STANTON, R. E., MCDONALD, A. J. and CARMICHEL, I.: *Analyst* **87**, 134 (1962).
442. STARÝ, J.: *Anal. Chim. Acta* **28**, 132 (1963).
443. STARÝ, J. and HLADKY, E.: *Anal. Chim. Acta* **28**, 227 (1963).
444. STARÝ, J. and RUŽIČKA, J.: *Talanta* **8**, 775 (1961).
445. STATEN, F. W. and HUFFMAN, E. W. D.: *Anal. Chem.* **31**, 2003 (1959).
446. STEINBACH, J. F. and FREISER, H.: *Anal. Chem.* **25**, 881 (1953).
447. STEINBACH, J. F. and FREISER, H.: *Anal. Chem.* **26**, 375 (1954).
448. STOLYAROV, K. P.: *Zh. Anal. Khim.* **16**, 452 (1961).
449. STROMBERG, A. and ZELJANSKAJA, A. I.: *Zh. obshc. Khim.* **15**, 303 (1945).
450. SUBBARAMAN, P. R., CORDES, SR. M. and FREISER,hH.: *Anal. Chem.* **41**, 1848 (1969).
451. SUZUKI, N. and OMORI, T.: *Bull. Chem. Soc. Japan* **35**, 595 (1962).
452. SVOBODA, V.: *Quercetin.* Chemapol. **14**. Prague (1959).
453. SZABÓ, Z. G. and BECK, M. T.: *Anal. Chem.* **25**, 103 (1953).
454. SZABÓ, Z. G., BURGER, K. and KŐRÖS, E.: *Kémiai Közlemények* **30**, 245 (1968).
455. TAKIYAMA, K., SALESIN, E. D. and GORDON, L.: *Talanta* **5**, 231 (1960).
456. TOMICEK, O., SPURNY, K., JERMAN, L. and HOLECEK, V.: *Coll. Czech. Chem. Commun.* **18**, 757 (1953).
457. TOUGARINOFF, M. B.: *Ann. Soc. Sci. Bruxelles* **54B**, 314 (1934).
458. TSCHUGAEFF, L. Z.: *Z. anorg. Chem.* **46**, 144 (1905).
459. TSUCHIDA, R.: *Bull. Chem. Soc. Japan* **13**, 388, 436, 471 (1938).
460. TSUCHIDA, R. and KOBAYASHI, M.: *Bull. Chem. Soc. Japan* **13**, 47 (1938).
460a. UMLAND, F.: *Theorie und praktische Anwendung von Komplexbildnern.* Akademische Verlagsges., Frankfurt am Main, 1971.
461. UMLAND, F., HOFFMANN, W.: *Z. anal. Chem.* **168**, 268 (1959).
462. UMLAND, F., HOFFMANN, W. and MECKENSTOCK, K.: *Z. anal. Chem.* **185**, 362 (1962).
463. UPOR, E.: *Hidrológiai Közlöny* **38**, 299 (1958).
464. VAN KLOOSTER, H. S.: *J. Amer. Chem. Soc.* **93**, 746 (1921).
465. VAN UITERT, L. G. and FERNELIUS, W. C.: *J. Amer. Chem. Soc.* **76**, 375 (1954).
466. VERDIER, E. T.: *Coll. Czech. Chem. Comm.* **11**, 233 (1939).
467. WADE, M. A. and SEIM, H. J.: *Anal. Chem.* **33**, 793 (1961).
468. WAKAMATSU, S.: *Japan Analyst (Bunseki Kagaku)* **9**, 284 (1960).
469. WAWRZYCZEK, W. and MAJKOWSKA, H.: *Z. anal. Chem.* **199**, 430 (1964).
470. WEHLER, P.: *Z. anal. Chem.* **153**, 249 (1956).
471. WEISS, H. V. and LAI, M. G.: *Anal. Chem.* **32**, 475 (1960).
472. WEISS, H. V., LAI, M. G. and GILLESPIE, A.: *Anal. Chim. Acta* **25**, 550 (1961).
473. WELCHER, F. J.: *Organic Analytical Reagents.* D. Van Nostrand Co., New York, 1947.
474. WELCHER, F. J.: *Analytical Chemistry.* Proc. Int. Symp. Birmingham, 1962, p. 106.
475. WENDLANDT, W. W.: *Anal. Chim. Acta* **16**, 216 (1957).
476. WENGER, P. and MASSET, E.: *Helv. Chim. Acta* **23**, 38 (1940).
477. WIERSMA, L. D. and LOTT, P. F.: *Anal. Chim. Acta* **40**, 291 (1968).
478. WILLARD, H. H. and DEAN, J. A.: *Anal. Chem.* **22**, 1264 (1960).
479. WILLIAMS, R. J. P.: *Analyst* **78**, 586 (1953).

480. WILLIAMS, D. E., WOHLAUER, G. and RUNDLE, R. E.: *J. Amer. Chem. Soc.* **81**, 755 (1959).
481. WÖLBLING, H. and STEIGER, B.: *Mikrochemie* **15**, 295 (1934).
482. YASUDA, M., SONE, K. and YAMASAKI, K.: *J. Phys. Chem.* **60**, 1667 (1956).
483. YATSIMIRSKII, K. B.: *Termokhimiya Kompleksnikh Soedinenii*, Izd. Akad. Nauk SSSR, Moscow, 1951.
484. YATSIMIRSKII, K. B.: *Zh. Anal. Khim.* **10**, 94 (1955).
485. YATSIMIRSKII, K. B.: *Zh. Neorg. Khim.* **2**, 2346 (1957).
486. YATSIMIRSKII, K. B. and VASILOV, I.: *Instability Constants of Complex Compounds* Pergamon Press, Oxford, 1960.
487. YOE, J. H.: *J. Amer. Chem. Soc.* **54**, 4139 (1932).
488. YOE, J. H. and HALL, R. T.: *J. Amer. Chem. Soc.* **59**, 872 (1937).
489. YOUNG, R. S.: *Analyst* **76**, 49 (1951).
490. ZANKO, A. M. and BUTENKO, G. A.: *Zav. Lab.* **6**, 545 (1937).
491. ZANKO, A. M. and BUTENKO, G. A.: *Dokl. Akad. Nauk SSSR* **9**, 99 (1938).
492. ZHAROVSKII, F. G. and RIZHENKO, V. L.: *Ukr. Khim. Zh.* **28**, 306 (1962).
493. ZOLOTOV, YU. A.: *Acta Chim. Acad. Sci. Hung.* **32**, 327 (1962).
494. ZOLOTOV, YU. A.: *Ekstraktsiya Vnutrikompleksnikh Soedinenii*, Izd. Nauk SSSR, Moscow, 1968.
495. ZOLOTOV, YU. A., SHAKLOVA, N. V. and ALIMARIN, I. P.: *Zh. Anal. Khim.* **23**, 1321 (1968).
496. ZOLOTOV, YU. A. and VLASOVA, G. E.: *Zh. Anal. Khim.* **24**, 1542 (1969).

AUTHOR INDEX

ABBOT, C. D. *123*, 239
ABRAHAMSON, E. W. *69*, 247
ACKERMANN, H. *20*, 214, 219, 247
ADAM, J. A. *115*, 239
AGREN, A. 228
AHRLAND, S. 16, 17, *24*, 239
ALIMARIN, I. P. 61, 70, *98*, *100*, *101*, *127*, *128*, 137, 239, 240, 243, 248, 249
ALMÁSY, A. *78*, 239
ALMÁSSY, GY. *112*, 239, 244
ALTEN, F. *102*, 239
ANDEREGG, G. *20*, *21*, *23*, 212, 214, 219, 225, 226, 239, 247
ANDERSON, M. L. *117*, 244
ANDERSON, P. 220
ANDERSON, S. *25*, *61*, 216, 224, 239
ARAYA, S. *123*, 245
ARCHAKOVA, T. A. *136*, 243
ASADRA, M. *51*, 244
ASHBROOCK, A. W. *98*, 239
ASHIZAWA, T. *122*, *123*, 245
ASOP, P. A. *119*, 244
ASTAHOVA, E. K. *89*, 246
ATHAVALE, V. T. *123*, 239
AYRES, G. H. *66*, *130*, 239

BABKO, A. K. *98*, *112*, *122*, 239
BAES, C. F. *74*, 240
BAGINSKI, E. S. *110*, 243
BAGREJEV, V. V. *136*, 247
BAILEY, T. H. 239
BAMBACH, K. *70*, *123*, 239
BANKS, C. V. *12*, *25*, 61, 216, 224, 226, 228, 239, 245
BÁNYAI, E. 239
BARBER, H. H. *132*, 239
BARD, J. A. 239
BARDI, R. 242
BARNES, H. *123*, 239
BARNES, J. W. 239
BARNUM, D. W. *61*, 239
BASOLO, F. *24*, 239
BASSET, J. *88*, 239
BATES, R. G. 229
BAYER, E. 13, 239
BAZHANOVA, L. A. *122*, 241
BEAUFAIT, L. J. *117*, 243
BECK, G. *111*, 239

BECK, M. T. *12*, 13, *26*, *207*, 208, 214, 219, 239, 248
BELCHER, R. 38, 39, *72*, 82, 92, 115, *138*, 239, 240
BELL, C. F. *119*, *122*, *123*, *129*, 243
BERG, R. 56, *72*, *96*, *102*, *130*, 240
BERGER, W. *106*, *107*, 240
BERSIN, TH. *131*, 240
BERTSCH, C. R. 220
BETTERIDGE, D. *38*, 240
BETTS, R. H. *117*, 240
BEZZI, S. 240, 242
BHATKI, K. S. *53*, 245
BHATNAGER, D. C. *71*, 244
BILLO, E. J. *69*, 240
BIRYUKOV, A. A. *61*, 239, 240, 248
BJERRUM, J. *19*, 23, 208, 220, 240
BLAEDEL, W. J. 78, 240
BLAKE, C. A. *74*, 240
BLANARU, E. *82*, 246
BLAU, F. 240
BLINC, R. *59*, 240
BLOCK, B. P. 211, 213, 220
BOCHKOVA, V. M. *59*, *89*, *209*, 216, 224, 240, 246
BODE, H. *124*, 125, *126*, *127*, *128*, 240
BOGDANOVICH, L. I. *107*, 241
BOHIGIAN, T. A. 225
BOLOMEY, R. A. *117*, 240
BOLTZ, T. F. *65*, *108*, 243
BOOTH, E. *115*, 239
BORDNER, J. *69*, 240
BORUFF, C. S. *102*, 243
BOSE, M. K. *109*, 246
BRAININA, K. Z. *78*, 240
BRANDŠTETR, I. *114*, 240
BRANNAN, J. R. *79*, 240
BRAY, R. H. *105*, *106*, *125*, *126*, 241
BREANT, M. *123*, 240
BREWER, J. G. *70*, 242
BRISCOE, G. B. 119, 240
BRITO, F. 226
BRITTEN, E. F. 231
BROTZ, W. *137*, 246
BROWN, K. B. *74*, 240
BROWN, W. B. *114*, 240
BRUNISHOLZ, G. *79*, 245
BRYAN, R. F. *119*, 240

Italic page numbers inicate a citation by reference number only.

BUA, E. 240
BUDESINSKY, B. *135*, 240
BULLWINKEL, E. P. 240
BURGER, K. *12*, 13, *16*, *17*, *18*, *25*, 41, *42*, *45*, 46, *48*, *51*, *59*, 61, *62*, *71*, *76*, *77*, *83*, 84, *87*, *89*, *92*, *209*, 213, 215, 216, 217, 220, 222, 224, 227, 228, 230, 240, 241, 246, 248
BURSTALL, F. *58*, 245
BUSEV, A. I. *11*, *107*, *122*, 124, 219, 241
BUTENKO, G. A. *109*, 249
BUTYLKIN, L. P. 128, 244
BUZÁS, I. *137*, 242
BYRD, C. H. *12*, 245

CAGLE, F. W. *22*, 241
CALLAHAN, C. M. 213
CALVIN, M. 13, *116*, 241, 245, 246
CAMPBELL, M. E. *99*, *100*, *101*, 244
CANHAM, A. G. 229
CARMICHEL, I. *122*, *123*, 248
CHABEREK, S. 225
CHAN, F. L. *70*, 241
CHANDELLE, R. 133, 241
CHARLAMOV, I. P. *124*, 241
CHARLES, R. G. *59*, *209*, 241
CHATT, J. 16, *17*, *24*, *42*, 239, 241
CHENG, K. L. *12*, 36, 37, *39*, *105*, *106*, *107*, *125*, *126*, 241
CHERNIKOV, YU. A. *98*, *134*, 241
CHERNITSKAYA, R. E. *99*, *100*, *101*, 241
CHOPPIN, R. G. *78*, 241
CLAASSEN, R. I. *102*, 243
CLARKE, K. *18*, 241
CLAYTON, R. F. *101*, 241
CLIFFORD, P. A. *123*, 241
CLUETT, M. L. *126*, 241
COHEN, A. J. *100*, *101*, 245
COHEN, J. *123*, 246
COLEMAN, C. F. *74*, 240
CONNICK, R. E. *118*, 241
COOK, E. H. *116*, 241
COOKSEY, B. G. *119*, 240
COOPER, S. S. *123*, 241
CORDES, S. M. 248
CORKINS, J. T. *69*, 241
CORSINI, A. *17*, 242
COTTON, F. A. *115*, 244
COURTNEY, R. C. 225
COWEN, R. A. *18*, 241
CRIVAT, D. *82*, 246
CYRANKOWSKA, M. *122*, 242

DAHLBERG, V. 242
DALE, J. M. 226
DAVAADORZH, P. *127*, 243
DAVIES, N. R. 16, *17*, *24*, 239
DAVIS, W. F. *83*, 242
DAWIDSON, J. E. *130*, 247
DAWSON, E. C. *122*, 242

DAY, A. R. *72*, 242
DE, A. K. *117*, *118*, 242, 244
DEAN, J. A. 77, *117*, 242, 248
DE BRUIN, H. J. *117*, 242
DEIJS, W. B. *122*, 243
DEWEY, D. W. *121*, 245
DINSTL, G. *126*, 242
DITT, M. *65*, *108*, 243
DOBKINA, B. M. *98*, 241
DOROUGH, G. D. 239
DOUGLAS, B. E. 230
DROLL, H. A. 211
DUMITRU, M. *81*, 246
DUVAL, C. 205, 242
DYRSSEN, D. *13*, *22*, *51*, *59*, 61, *100*, *101*, *209*, 213, 216, 222, 241, 242
DYRSSEN, M. 213

EARING, M. H. *125*, 244
ECKERT, H. W. *123*, 242
EDWARDS, L. J. 220
EGYED, I. *18*, *25*, *62*, *89*, 227, 228, 241
ELLEFSEN, P. R. *69*, 243
ELLIOT, G. R. *122*, 242
ELVERS, H. *106*, *107*, 240
EPHRAIM, F. 62, 242
ERDEY, L. *13*, *107*, *122*, *137*, 205, 242, 247
ERDEYNÉ, SCHNEER, A. *65*, *108*, *112*, 242
EREMINA, G. V. *128*, 243
ESHELMAN, H. C. *117*, 242
EVERETT, M. E. *123*, 246

FAIRMAN, W. O. *117*, 245
FAJANS, K. *42*, 44, 242
FARSANG, GY. *76*, *77*, *87*, 241
FEIGL, F. 11, 12, 13, 58, 61, *62*, 63, 64, 66, *93*, 242, 245
FELDMEIJER, J. H. *122*, 243
FERNANDO, Q. *17*, *25*, *26*, *49*, 242, 245, 247
FERNELIUS, W. C. *59*, *209*, 211, 220, 229, 230, 248
FERRUS, R. 225
FIRSCHING, F. H. 70, 242
FISCHER, H. *122*, 242
FLAGG, J. F. 83, 84, *89*, 242, 243
FLASCHKA, H. *39*, 242
FLECK, G. M. 242
FLEPS, V. *122*, 242
FOGG, H. C. *133*, 246
FORBLOM, S. *79*, 247
FOREMAN, B. M. *117*, *118*, 246
FORSTER, W. A. *126*, 242
FORTUNE, W. B. *80*, 242
FRASSON, E. 242, 246
FREASIER, B. F. 223
FREISER, B. S. 119, 242
FREISER, H. 13, 17, *25*, *26*, 49, *53*, *59*, *74*, *114*, *115*, 119, *124*, *207*, 222, 241, 242, 245, 247, 248

AUTHOR INDEX

FREITAG, E. 219, 225, 247
FRIEDEBERG, H. *120*, *128*, 242
FRONEAUS, S. 229
FUJITA, J. *59*, 243, 245
FUNK, H. *65*, *108*, 243
FURMANN, N. H. 83, 84, *89*, 242, 243
FYFE, W. S. 231

GAHLER, A. R. *82*, 243
GAIZER, F. 84, 241
GALLEGO, R. *122*, 243
GANCHOF, J. G. *117*, 245
GARDNER, K. *100*, *101*, 243
GEDDA, K. *126*, 245
GENDLER, S. M. 229
GENTRY, C. H. R. *98*, *99*, *100*, 243
GERBER, L. *102*, 243
GIBALO, I. M. *70*, *127*, *128*, 239, 243
GILLARD, R. D. *59*, 243
GILLESPIE, A. *70*, 248
GODYCKI, L. E. 243
GOLDSTEIN, G. *106*, 243
GOLOVINA, A. P. *98*, *100*, *101*, 239
GORDON, L. *69*, 240, 241, 243, 244, 245, 247, 248
GORJUSINA, V. G. *136*, 243
GÖRÖG, S. 214, 215, 219, 239
GRAHAM, R. P. *69*, 240
GRAY, G. W. *18*, 241
GRENTHE, J. 211
GRÖSSEL, V. G. 241
GRZESKOVIAK, E. *132*, 239
GUPTA, J. *109*, 246
GUSTAFSON, R. L. 223, 225, 231
GUT, R. 213, 219, 247
GYENES, I. 243

HAAS, C. G. 211
HADDOCK, L. A. *57*, *121*, 243
HADŽI, D. *59*, 240
HAGEMANN, F. *116*, *117*, *118*, 243
HAHN, R. B. *110*, 243
HALL, A. J. *85*, 243
HALL, R. T. *56*, 249
HALMOS, T. *111*, 247
HAMMES, G. H. 49, 243
HANANIA, G. I. H. 227
HARDWICK, W. H. *101*, 241
HARGIS, L. G. *65*, *108*, 243
HARTMANN, H. *44*, *65*, *71*, *72*, *110*, *111*, *137*, 243, 247
HASHITANI, H. *99*, *100*, *101*, 243, 245
HECHT, F. *126*, 242
HELLER, J. 219, 225
HENNICHS, M. *59*, *209*, 216, 242
HENRIQUES, R. *11*, 243
HEYNDRICK, A. 224
HILEMAN, O. E. *69*, 243, 244
HINSWARK, O. N. *76*, 243

HLADKY, E. *114*, 248
HOFFMANN, J. L. *123*, 245
HOFFMANN, W. *98*, *99*, *100*, *101*, 248
HOLECEK, V. *112*, 248
HOLLOWAY, J. H. 214
HOLZER, H. *89*, 243
HOPFF, H. *113*, 245
HORI, T. *122*, 243
HOUFF, W. H. *76*, 243
HOWICK, L. C. *70*, 244
HSEN, T. M. 214
HUBBARD, D. M. *122*, 243
HUDGENS, J. E. *117*, 245
HUFFMAN, E. H. *117*, 243
HUFFMAN, E. W. D. *107*, 248
HULANICKI, A. 129, 243
HYNEK, R. J. *105*, 243

IGNATEVA, N. G. *61*, 246
ILINSKI, J. *11*, 243
INCZÉDY, J. 79, 243
INTORRE, B. J. 225
IRVINE, D. H. 227
IRVING, H. *12*, 13, 17, *21*, *22*, *58*, *117*, 119, *122*, *123*, *129*, 221, 222, 243
ISHIHARA, Y. *122*, 243
IVANOV, V. M. *107*, 241
IWANTSCHEFF, G. *122*, *123*, 244
IZATT, R. M. 211

JAGUER, D. 242
JAMES, C. *133*, 246
JASIM, F. *100*, *101*, 244
JENIK, J. *41*, *126*, 246
JENKINS, C. R. *82*, *115*, **240**
JENNINGS, J. *63*, 244
JERMAN, L. *112*, 248
JILLOT, B. A. *61*, 244
JOHANSSON, E. 213
JOHNSON, E. I. *123*, 239
JOHNSTON, W. D. 222
JONES, D. L. *49*, 245
JONES, J. G. *18*, 244
JONES, J. L. *70*, 244
JONES, J. P. *69*, 244
JØRGENSEN, C. K. *46*, 244
JUSTUS, N. L. 219

KALINKIN, P. *126*, 244
KANNER, L. J. *69*, 244
KATO, M. *21*, 248
KATO, S. 247
KAWAHATA, M. *122*, *123*, 244
KEENAN, T. K. *117*, 230, 244
KHOPKAR, S. M. *117*, *118*, 244
KIMURA, K. *51*, *126*, 244
KIRSCHNER, S. 71, 244
KIRSON, B. 227
KISELEVA, L. V. *107*, 241
KISHI, H. *122*, 243

Kiss, A. *112*, 244
Klement, R. *95*, 244
Knopf, P. M. *119*, 240
Knowles, H. 63, 244
Kobayashi, M. *44*, 248
Kobrová, M. *126*, 246
Koch, O. G. *126*, 245
Kolarik, Z. *116*, *117*, 244
Kolthoff, I. M. 224
Korecz, L. *13*, *46*, *60*, 215, 220, 230, 241
Korenman, I. M. *12*, 244
Koroleff, F. *118*, *122*, *123*, 244
Körös, E. *17*, 248
Kozhokina, G. J. A. 247
Krapivkina, T. A. *78*, 240
Krasovec, F. *22*, 216, 242
Krausz, I. 97, 137, 239, 244
Kreimer, S. E. 128, 244
Krishen, A. *114*, 244
Kuljberg, L. M. *11*, *12*, *13*, 244
Kumins, Ch. A. 65, *110*, 244
Kuta, J. *76*, 246
Kuznetsov, V. I. *12*, *134*, *136*, 244

Lacoste, R. J. *125*, 244
Lai, M. G. *70*, 244, 248
Laing, M. *119*, 244
Lane, T. J. 231
Laszlovszky, J. *122*, 247
Legzdins, P. *115*, 244
Leigh, R. M. *117*, 240
Leopoldi, G. *122*, 242
Lerner, M. *78*, 244
Letow, T. B. *88*, 239
Leussing, L. D. 224
Lewis, J. *25*, *26*, *51*, 244
Lindenbaum, A. 229
Lindner, M. *118*, 244
Lippard, S. J. 115, 244
Loofman, H. *102*, 239
Lott, P. F. *88*, 248
Luke, C. L. *99*, *100*, *101*, 244
Lukyanov, V. F. *136*, 244
Lumme, P. O. *64*, 244
Lunden, L. *79*, 247
Lyle, S. J. *70*, 239, 244

Mag, P. *13*, *60*, 241
Magee, R. J. *69*, *100*, *101*, 243, 244
Magnusson, L. B. *117*, 244
Mairanovskii, S. G. *87*, 244
Maissen, B. *20*, 247
Majer, J. *115*, 240
Majkowska, H. *130*, 248
Majumdar, A. K. *117*, 245
Malissa, H. *126*, *127*, 245
Mammi, M. 242
Manning, D. L. *106*, 243
Manuaba, I. B. A. *13*, *46*, *60*, 241

Marcus, Y. *23*, 245
Marec, D. J. *69*, 245
Margerum, D. W. *12*, 49, 245
Marston, H. R. *121*, 245
Martell, A. E. *17*, *27*, *207*, 219, 223, 225, 230, 231, 248
Masset, E. *108*, 248
Math, K. S. *17*, *53*, 245
Maverick, E. F. 219
May, I. *123*, 245
Mayr, C. 58, *93*, *95*, 245
McDonald, A. J. *122*, *123*, 248
McIntyre, G. H. 220
McKaveney, J. P. *114*, *115*, 245
McVey, W. H. *118*, 241
Meckenstock, K. *98*, *100*, *101*, 248
Meek, D. W. *115*, 248
Meinke, W. W. *118*, 247
Meites, L. *75*, *77*, *78*, *87*, 245, 248
Mellon, M. G. *80*, *82*, 242, 245
Mellor, D. H. *21*, *58*, 243
Melsted, S. W. *125*, *126*, 241
Menis, O. *106*, *117*, 242, 243
Merritt, L. L. *105*, 245
Meyer, K. H. *113*, 245
Meyer, S. *126*, 245
Miller, F. F. *126*, 245
Miller, W. L. *100*, *101*, *123*, 245
Miranda, L. I. *58*, *61*, 242
Misaki, T. *122*, *123*, 244
Miyahara, K. *122*, 245
Miyamoto, M. *123*, 245
Mochizuki, M. *122*, *123*, 244
Moeller, T. *99*, *100*, *101*, 214, 225, 245
Moore, F. L. *117*, *118*, 245
Moreton-Smith, M. *101*, 241
Moret, R. *79*, 245
Morgan, G. 58, 245
Morgen, L. O. 219
Morimoto, Y. *122*, *123*, 245
Morris, M. L. *115*, 248
Morrison, G. H. *49*, *74*, *114*, *115*, 242, 245
Moser, P. *19*, 247
Moshier, R. W. *115*, 245, 248
Moss, M. L. *82*, 245
Motojima, K. *99*, *100*, *101*, 245
Muchina, Z. S. *122*, 245
Muk, A. A. *136*, 247
Murakami, Y. *126*, 244
Murmann, R. K. *24*, 239
Muxard, R. *117*, 245
Myassoedov, B. *117*, 244, 245

Nakahara, A. *59*, 243, 245
Nance, K. V. *90*, 245
Nancollas, G. H. *79*, 240
Narang, B. D. *66*, 239
Näsänen, R. *22*, 222, 223, 231, 245
Nazarenko, V. A. *126*, 245

NEDELJAK, M. 95, 245
NELSON, E. J. 220
NEUMANN, F. 240
NI, G. V. 244
NIELSCH, W. 85, 245
NIKOLSKAYA, I. B. 136, 244
NOBLE, P. 240
NORWITZ, G. 123, 246
NOVAK, J. V. A. 76, 246

OBERG, A. G. 223
OKAČ, A. 111, 246
OKHANOVA, L. A. 244
OLSEN, E. D. 78, 240
OMORI, T. 114, 248
ORDELT, H. 63, 242
ORGEL, L. E. 16, 17, 42, 246
OSBORNE, E. H. 18, 241

PALEJ, P. N. 134, 246
PALMA, R. J. 71, 246
PANATTONI, C. 242, 246
PÁNKOVA, H. 116, 117, 244
PAPP-MOLNÁR, E. 212, 215, 217, 220, 222, 224, 227, 241
PARRY, R. W. 20, 248
PATOH, P. 111, 247
PAULIK, J. 77, 246
PAULING, L. 17, 44, 246
PEARSON, K. H. 71, 246
PEARSON, R. G. 17, 246
PECSOK, R. L. 219
PECZOK, P. L. 220
PERKINS, M. 77, 246
PERLICH, R. W. 121, 247
PERRIN, D. D. 11, 64, 228, 246
PERRY, R. 115, 240
PESHKOVA, V. M. 59, 61, 89, 114, 209, 216, 224, 240, 246
PETERSON, H. E. 123, 248
PETERSON, S. 116, 246
PETRUKIN, O. M. 101, 239
PETTIT, L. 21, 58, 243
PEVZNER, N. S. 134, 241
PIETRZAK, P. F. 69, 241
PIGAGA, A. K. 70, 239
PILIPENKO, A. T. 122, 123, 239, 246
PINTÉR, B. 48, 61, 241
PIRTEA, T. I. 81, 82, 246
PODTZAYNOVA, V. H. 126, 246
POLIANSKII, N. G. 11, 241
POLUEKTOFF, N. S. 66, 246
PONDER, B. W. 115, 248
POOL, J. B. 18, 244
POSKANZER, A. M. 117, 118, 246
PRESTON, P. F. 122, 242
PŘIBIL, R. 41, 71, 126, 246
PROSZT, J. 77, 246
PRUE, J. 19, 20, 246

PRZHEVALSKII, E. S. 98, 100, 101, 239
PUZDRENKOVA, I. V. 98, 100, 101, 239
PYATNICKY, I. V. 229

RABINOWITCH, E. 23, 246
RÁDY, G. 122, 242
RAFF, P. 137, 246
RAHAMAN, M. S. 117, 242
RAINS, T. S. 117, 242
RÄKER, K. O. 122, 246
RAMACHANDRAN, T. P. 123, 239
RAMAKRISHNA, R. S. 122, 243
RANDOLPH, D. R. 117, 247
RAO, B. S. K. 246
RAO, B. S. V. R. 246
RAY, P. 109, 112, 130, 246
REED, S. A. 12, 245
REID, J. C. 116, 246
REILLY, C. N. 214
REINBOLD, P. E. 71, 246
REYNOLDS, G. F. 77, 246
RICE, A. C. 133, 246
RICHARD, C. 223, 231
RIEMAN, W. 78, 244
RIEBOLD, P. E. 71, 246
RIHA, J. 76, 246
RINGBOM, A. 12, 13, 27, 28, 36, 219, 246, 247
RITCEY, G. M. 98, 239
RIZHENKO, V. L. 99, 249
ROBERTSON, B. E. 69, 240
ROEBLING, W. 130, 240
ROLLET, P. 61, 247
ROMANOVA, E. V. 136, 243
ROONEY, R. C. 98, 247
ROSSOTTI, H. S. 12, 22, 221, 228, 243
RUDENKO, N. P. 114, 247
RÜDORF, W. 137, 247
RUFF, F. 13, 59, 61, 87, 89, 241
RUFF, I. 12, 13, 25, 48, 51, 59, 61, 87, 89, 216, 224, 227, 241
RUNDLE, R. E. 243, 248
RUZIČKA, J. 114, 248
RYAN, J. A. 231
RYDBERG, B. 114, 247
RYDBERG, J. 113, 114, 211, 247

SACHDEV, S. L. 128, 247
SAITO, K. 51, 244
SAJÓ, I. 71, 219, 247
SALESIN, E. D. 69, 240, 244, 245, 247, 248,
SAMUELSON, O. 79, 247
SANDELL, E. B. 56, 81, 101, 119, 121, 247
SANDERA, J. 219
SAPOZHNIKOVA, E. YA. 78, 240
SARMA, D. V. N. 246
SAVVIN, S. B. 134, 135, 136, 244, 247
SAWADA, T. 247
SCHIAVINATO, G. 240
SCHNEER, A. 65, 71, 72, 107, 110, 111, 112, 137, 247

SCHNEIDER, W. 225
SCHRAMM, K. 79, 247
SCHUBERT, J. 229
SCHULEK, E. 84, *122*, 241, 247
SCHWARZENBACH, G. *12*, *19*, *20*, 27, 29, *71*, 214, 219, 225, 231, 246, 247
SCHWEITZER, G. K. *117*, *130*, 247
SEIM, H. J. *78*, 248
SEKIDO, E. *17*, 247
SELL, H. M. *76*, 243
SEMIKOZOV, G. S. *126*, 244
SENN, H. 225
SEYANOVA, F. R. 247
SHAKLOVA, N. V. 249
SHEPERD, E. *118*, 247
SHERRINGTON, L. G. *98*, *99*, *100*, 243
SHIBATA, K. *122*, 243
SHIBATA, S. *105*, *106*, *107*, 247
SHIGEMATSU, T. *114*, 248
SHITAREVA, G. G. *126*, 245
SHLENSKAYA, V. I. *61*, *89*, 239, 240, 246, 248
SHTOKALO, M. I. *112*, 239
SICHER, G. *62*, 242
SIDGWICK, N. V. 15, 248
SIEVERS, R. E. *115*, 245, 248
SILL, C. W. *123*, 248
SILLÉN, L. G. *17*, *22*, *27*, *207*, 216, 219, 230, 242, 248
SILVA, R. J. *78*, 241
SIMION-NICOLAU, J. *81*, 246
SINGER, O. *62*, 242
SINGH, R. S. 228
SINYAKOVA, S. I. *122*, 248
SJÖSTRÖM, E. *79*, 247
SKRIFVARS, B. 219
SMIRNOVA, O. W. *124*, 241
SMITH, G. F. *22*, *94*, 241, 248
SNYDER, L. J. *121*, 248
SOKOLOVA, T. A. 219
SOKOLOV, S. S. *89*, 246
SONE, K. *21*, 248, 249
SOUTHERN, D. L. *70*, 244
SPIKE, C. G. 20, 248
SPITZER, M. *77*, *87*, 248
SPRINGER, C. S. *115*, 248
SPURNY, K. *112*, 248
STANTON, R. E. *122*, *123*, 248
STARÝ, J. *98*, *99*, *100*, *101*, 113, *114*, 247, 248
STATEN, F. W. *107*, 248
STEIGER, B. *130*, 248
STEINBACH, J. F. *114*, 240, 248
STENFELD, V. I. *49*, 243
STEPHEN, W. I. *82*, *92*, *115*, 240
STOCKMAYER, W. H. *23*, 246
STOLYAROV, K. P. *126*, 248
STRICKLAND, J. D. H. *115*, 239
STRICKS, W. 224
STROMBERG, A. *76*, 248
SUBBARAMAN, P. R. 248

SULLIVAN, M. L. *123*, 241
SURAK, J. G. *117*, 245
SÜTONEN, S. 219
SUTTLE, J. F. *117*, 230, 244
SUZUKI, N. *114*, 248
SVOBODA, V. *112*, 248
SYREK, G. *76*, *77*, *87*, 241
SZABÓ, Z. G. *12*, 13, *17*, *26*, 248
SZÁNTÓ-HORVÁTH, G. 217, 241

TABUSHI, M. *114*, 248
TAFT, R. V. *116*, 241
TAGGART, W. T. *72*, 242
TAKIYAMA, K. *69*, 248
TEMPLE, R. B. *117*, 242
THOMPSON, J. H. *122*, 242
TIKHONOVA, A. A. *122*, 245
TIKHVINSKAYA, T. I. *61*, 239, 240, 248
TILLU, M. M. *123*, 239
TIPDIVA, V. G. 219
TODD, R. *101*, 241
TOMICEK, O. *112*, 248
TOMKINSON, J. C. *18*, 221, 222, 223, 244
TÓTH, A. 230
TÓTH, T. *65*, *108*, 242
TOUGARINOFF, M. B. 83, 248
TOWNSEND, A. *69*, 244
TRAMM, R. S. *134*, 241
TRUJILLO, R. 226
TSCHUGAEFF, L. *11*, 248
TSUCHIDA, D. *59*, 243
TSUCHIDA, R. *42*, 44, *59*, 245, 248
TUSCHE, K. J. *124*, *125*, 240
TVESTKOVA, L. A. *122*, 248

UDEN, P. C. *82*, *115*, 240
UISITALO, E. 223
UMLAND, F. *98*, *99*, *100*, *101*, 248
UPOR, E. *112*, 248

VAIDYA, G. M. *123*, 239
VAN KLOOSTER, H. S. 57, 248
VAN UITERT, L. G. *59*, *209*, 229, 230, 248
VÁSÁRHELYI-NAGY, H. 215, 220, 227
VASILOV, I. 249
VEIVO, J. 231
VERDIER, E. T. *76*, 248
VIGH, K. *137*, 242
VIGVÁRI, M. 239
VLASOVA, G. E. 249
VOGEL, A. I. *88*, 239
VON USLAR, H. *122*, 242
VREŠTAL, J. *114*, 240

WACHTER, L. E. *100*, *101*, *123*, 245
WADE, M. A. *78*, 248
WAGNER, W. F. *114*, 240
WAKAMATSU, S. *107*, 248
WALKER, J. K. *105*, 245

Walter, J. L. 231
Wänninen, E. *13*, *28*, 247
Wawrzyczek, W. *130*, 248
Wehler, P. 27, *98*, 248
Weiland, H. *102*, 239
Weiss, H. V. *70*, 244, 248
Welcher, F. J. *11*, *13*, 79, 248
Wendlandt, W. W. 137, 223, 248
Wengelin, F. 242
Wenger, P. *108*, 248
Wert, P. W. *128*, 247
White, J. C. *74*, 240
Wiberly, S. E. *125*, 244
Wiersma, L. D. *88*, 248
Wilkins, R. G. *25*, *26*, *51*, 244
Wilkinson, G. *59*, 243
Willard, H. H. 77, 248
Willi, A. 231
Williams, A. A. *24*, 239
Williams, D. E. 248
Williams, R. J. P. *12*, 13, 17, 18, *21*, *58*, *61*, *122*, *123*, *129*, 221, 222, 223, 243, 244, 248
Wilson, C. L. *100*, *101*, 244
Wilson, K. W. *13*, 241
Wish, L. *117*, 240

Wittwer, S. H. *76*, 243
Wohlauer, G. 248
Wölbling, H. *130*, 248
Wrangell, L. J. *105*, 243

Yamamoto, K. *99*, *100*, *101*, 243
Yamasaki, K. *21*, 249
Yasuda, M. *21*, 249
Yatsimirskii, K. B. *12*, *16*, 27, *42*, 249
Yeremenko, O. M. 246
Yoe, J. H. *56*, 249
Yoe, J. M. *126*, 241
Yoshida, H. *99*, *100*, *101*, 245
Young, F. *130*, 239
Young, R. S. *85*, *123*, 243, 249

Zabin, B. A. *49*, 245
Zanetti, R. 242, 246
Zanko, A. M. *109*, 249
Zannier, H. *137*, 247
Zeljanskaya, A. I. *76*, 248
Zharovskii, F. G. *99*, 249
Zhemchuzhnaya, I. A. *122*, 245
Zolotov, J. A. *101*, *117*, 239, 249
Zozulya, A. P. *114*, 246

SUBJECT INDEX

Acetylacetone
  analytical application of 112–115
  extraction of metal ions with 114
  purification of, as reagent 113
  stability constants of its complexes 210
Actinium, extraction by TTA (thenoyltrifluoroacetone) 116, 117
Acyloinoxime 62–63
Aluminium
  detection by oxine 56
  determination by
    gravimetry using
      8-hydroxyquinoline 102
      quercetin 112
    polarography 77
    solvent extraction using
      acetylacetone 114
      8-hydroxyquinoline 98, 102
      morin 111
      TTA 117
    spectrophotometry using
      aluminium 185
      Arsenazo III 135
      Eriochrome Cyanin 186
      8-hydroxyquinoline 98, 186
      sodium alizarin sulphonate 186
Ammine complexes
  formation equilibria 28–30
  stability constants of 19
Amplification methods 72, 137
Anthranilic acid
  analytical application of 64–65, 108–109
  solubility product of its complexes 64
Antimony
  determination with
    solvent extraction using diethyldithiocarbamate 125, 127
    spectrophotometry using
      Methyl Violet 198
      Rhodamine B 198
      thiourea 198
    selective reaction with pyrogallol 63–64
Arsanilic acid, analytical application of 133
Arsenazo I (Uranone), analytical application of 134
Arsenazo III, analytical application of 134–316

Back-coordination 17, 24, 42, 44, 45–47
Barium, determination by solvent extraction using
  8-hydroxyquinoline 98
  TTA (thenoyltrifluoroacetone) 117
Basic strength of ligand, effect on complex stabilities 18
Beryllium, determination by
  solvent extraction using
    acetylacetone 114
    8-hydroxyquinoline 98
    TTA (thenoyltrifluoroacetone) 117
  spectrophotometry using
    acetylacetone 187
    Arsenazo III 135
    Beryllon II 187
    Eriochrome Cyanin 187
    8-hydroxyquinoline 187
    4-(p-nitrophenylazo)-orcinol 187
    quinalizarin 187
    sulfosalicylic acid 188
Bismuth,
  determination by
    gravimetry using cyclohexane-1,2-dione dioxime 88
    solvent extraction using
      diethyldithiocarbamate 126
      dithizone 121, 122, 125
      8-hydroxyquinoline 98
      TTA (thenoyltrifluoroacetone) 117
    spectrometry using
      dithizone 121, 122, 188
      Komplexon III 188
      sodium diethyldithiocarbamate 125, 126, 128, 188
      thiourea 188
    selective reaction with pyrogallol 63–64
Bromine chloride standard solution 84

Cadmium, determination by
  gravimetry using anthranilic acid 108
  solvent extraction using
    dithizone 122
    8-hydroxyquinoline 98
    PAN (pyridylazonaphthol) 106
    TTA (thenoyltrifluoroacetone) 117

Cadmium, determination by (*continued*)
  spectrophotometry using
    Arsenazo III 135
    Cadion 188
    dithizone 189
    4-hydroxy-3-nitrophenylarsonic acid 189
    8-hydroxyquinoline 98
Calcium, determination by
  gravimetry using picrolonic acid 95
  polarography 78
  solvent extraction using
    8-hydroxyquinoline 98
    TTA (thenoyltrifluoroacetone) 117
  spectrophotometry using
    Arsenazo III 135
    murexide 188
    picrolonic acid 95
CDTA-complexes, stability constants of 214
Cerium, determination by
  solvent extraction using
    8-hydroxyquinoline 98
    TTA (thenoyltrifluoroacetone) 117
  spectrophotometry using
    Arsenazo III 135
    benzidine 189
    brucine 189
    8-hydroxyquinoline 189
Charge-transfer band 46, 52
Chelate effect 19–20
Chromium, determination by
  gravimetry using quercetin 112
  solvent extraction using
    acetylacetone 113, 114, 115
    8-hydroxyquinoline 99
    TTA (thenoyltrifluoroacetone) 117
  spectrophotometry using
    diphenylcarbazide 190
    8-hydroxyquinoline 99
    Komplexon III 189
Chromatography and ion-exchange 78–79, 97, 112
Cobalt
  detection by
    nitroso, R salt 57
    thioglycollic acid anilide 131
  determination by
    gravimetry using
      anthranilic acid 108
      2,2'-dipyridyl + SCN^- 82
      8-hydroxyquinoline 103
      α-nitroso-β-naphthol 92
      1-10-phenanthroline + SCN^- 81
    polarography 76
      in the presence of nickel 75, 87–88
      nitrosonaphthols 57–58, 69
    solvent extraction with
      acetylacetone 113, 115
      diethyldithiocarbamate 126
      dithizone 120, 121, 122
      8-hydroxyquinoline 99, 103
      α-nitroso-β-naphthol 92
      PAN (pyridylazonaphthol) 106
      TTA (thenoyltrifluoroacetone) 117
    spectrometry using
      Daxime
      dimethylglyoxime + I^- 48, 61, 85–86, 188
      dithizone 121, 122
      8-hydroxyquinoline 99, 103
      Komplexon III 191
      Komplexon III + H_2O_2 189
      1-nitroso-2-naphthol 189
      2-nitroso-1-naphthol 190
      nitroso R salt 190
      sodium diethyldithiocarbamate 126, 190
      2,2'2"-tripyridyl 190
  reaction with
    cupferron 69
Complexes
  light absorption of 51
  solubility of 49
  successive stability constants of 12
Conditional equilibrium constants 27–35
Copper,
  determination by
    gravimetry using
      anthranilic acid 108
      2,2'-dipyridyl + SCN^- 82
      8-hydroxyquinoline 103
      quinaldic acid 109, 110
      rubeanic acid 130
      salicylaldoxime 89–90
    solvent extraction with
      acetylacetone 114
      α-benzoinoxime 90
      diethyldithiocarbamate 126
      8-hydroxyquinoline 99, 103
      PAN (pyridylazonaphthol) 106
      TTA (thenoyltrifluoroacetone) 117
    spectrophotometry with
      2,2'-biquinoline 191
      2,3-bis[2-(-methyl)-pyridyl)] quinoxaline 82
      Daxime 91–92, 191
      2,9-dimethyl-4,7-diphenyl-1,10-phenanthroline 191
      1,3-dimethyl-4-imino-5-oxyiminoalloxane 91–92, 190
      2,9-dimethyl-1,10-phenanthroline (Neocuproine) 81–82
      dithizone 122, 191
      Komplexon III 191
      neocuproin 81–82
      nitrilotriacetic acid 191
      pyridine + SCN^- 191
      sodium diethyldithiocarbamate 191
      zinc dibenzyl dithiocarbamate 192

SUBJECT INDEX

Copper *(continued)*
  reaction with
    anthranilic acid 64, 108
    dimethylglyoxime 59–61
    rubeanic acid 11, 130
    salicylaldoxime 62–63
Coprecipitation 68, 70
Cupferron, analytical application of 68, 94
Cuproine group, specific character of 21, 58, 81–82

Daxime (1,3-dimethylalloxane-imide(4)-oxime(5))
  analytical application of 91–92
  copper complex of 92
d-d Transition 45, 52
Demasking 35, 41
Diethylammonium diethyldithiocarbamate,
  analytical application of 128–129
p-Dimethylaminobenzylidenerhodanine,
  reactions of 65–66
Dimethylglyoxime
  analytical application of 11, 48, 59–61, 68, 83–88
  oxidimetric determination 84
Dimethylglyoxime complexes
  precipitation from homogeneous solution 69
  solubility of 61
  stability constants of 25, 215
  structure of 48, 59–61
2,9-Dimethyl-1,10-phenanthroline 21, 81–82
Dioximes, analytical application of 59–61, 69, 83–90
Dipicrylamine, analytical application of 66
2,2′-Dipyridyl
  analytical application of 58, 72, 82
  stability constants of its complexes 211
Distribution ratio 85
Dithiocarbamates
  analytical application of 68, 124–129
  stability constants of its complexes 215, 226
Dithizone (Diphenylthiocarbazone)
  analytical application of 118–124
  primary and secondary dithizonates 119
  purification 120–121
d-Orbitals
  energy diagram of 43
  spatial arrangement 42

EDTA complexes
  formation equilibria 27–31
  stability constants of 217–218
Electron transitions 51–53
  charge-transfer 52
  within the ligands 53
Equilibrium calculations 26–35
Ethylenediamine complexes
  stability constants of 19, 219
  use of, in polarography 78
Ethylxanthate complexes, stability constants of 219

Ferron (7-iodo-8-hydroxyquinoline-5-sulfonic acid) 56
Functional groups of organic reagents 54–56
α-Furyldioxime
  analytical application of 88–89
  stability constants of its complexes 212

Gallium, determination by
  solvent extraction using
    acetylacetone 114
    8-hydroxyquinoline 98
    PAN (pyridylazonaphthol) 106
  spectrophotometry using
    quinalizarin 193
    Rhodamine B 193
Gas chromatographic separations 115
Germanium, determination by spectrophotometry using phenylfluoron 193
Gold, determination by
  solvent extraction using dithizone 122
  spectrophotometry using
    p-dimethylaminobenzylidenerhodanine 187
    Rhodamine B 187
    o-toluidine 187

Hafnium, determination by
  gravimetry using phenylarsonic acid 132
  solvent extraction using TTA (thenoyltrifluoroacetone) 117
  spectrophotometry using
    acetylacetone 114
    Arsenazo III 135
Hard acids and bases 17
High-spin electronic structure 42, 44
8-Hydroxy-2-methylquinoline 50, 105
  stability constants of its complexes 220
8-Hydroxyquinoline
  analytical application of 56, 68, 69, 73–74, 96–104
  bromatometric determination of 72, 97, 104
  polarographic reduction of 78
  stability constants of its complexes 221
8-Hydroxyquinoline-5-sulfonic acid, stability constants of its complexes 222

Indium, determination by solvent extraction using
  acetylacetone 114
  5,7-dibromo-8-hydroxyquinoline 194
  dithizone 123
  8-hydroxyquinoline 98, 194
  PAN (pyridylazonaphthol) 106
  TTA (thenoyltrifluoroacetone) 117

Iridium, determination by
  solvent extraction using TTA (thenoyltrifluoroacetone) 117
  spectrometry using Komplexon III 194
Iron
  detection by 8-hydroxyquinoline 56
  determination by
    solvent extraction using
      acetylacetone 114, 115
      dithizone 122
      8-hydroxyquinoline 98
      PAN (pyridylazonaphthol) 106
      TTA (thenoyltrifluoroacetone) 117
    spectrophotometry using
      dimethylglyoxime 61, 192
      4,7-diphenyl-1,10-phenanthroline 192
      2,2'-dipyridyl 82, 192
      Ferron 56, 192
      Komplexon III 193
      Komplexon III + H_2O_2 193
      1,10-phenanthroline 80–81, 192
      salicylaldoxime 193
      salicylic acid 193
      sulfosalicylic acid 193
      tartaric acid 193
      2,2'2"-tripyridyl 192
    reaction with
      2,2'-dipyridyl 58
      1,10-phenanthroline 58

Kalignost (sodium tetraphenylborate), analytical application of 137–138

Lanthanum, determination by solvent extraction using
  Arsenazo III 135
  8-hydroxyquinoline 98
  TTA (thenoyltrifluoroacetone) 117
Lead, determination by
  solvent extraction using
    acetylacetone 114
    dithizone 121, 123, 196
    8-hydroxyquinoline 100
  spectrophotometry using
    Arsenazo III 135
    p-nitrosodiphenylamine 197
    1-nitroso-2-naphthol 197
    phenyl-α-pyridylketoxime 197
    tetramethyldiaminodiphenylmethane 197
Lithium, determination by spectrophotometry using Thoron 194
Low-spin electron structure 42, 44

Magnesium, determination by
  gravimetry using 8-hydroxyquinoline 104
  solvent extraction using 8-hydroxyquinoline 99, 195

  spectrophotometry using
    Arsenazo III 135
    Eriochrome Black T 195
    Titan Yellow 195
Mandelic acid (phenylglycollic acid), analytical application of 65, 110–111
Manganese, determination by
  gravimetry using anthranilic acid 108
  polarography 76
  solvent extraction using
    diethyldithiocarbamate 126
    8-hydroxyquinoline 100
    PAN (pyridylazonaphthol) 106
  spectrophotometry using Komplexon III 195
Masking
  agents 31–34, 39–41, 120, 125, 131
  factor (M. F.) 36–38
  ligands 39–41
    log α values of 31–34
  quantitative characterization of 36–38
Mercaptoacetic acid complexes, stability constants of 223
Mercury, determination by
  solvent extraction using
    dithizone 120, 123
    8-hydroxyquinoline 99
    PAN (pyridylazonaphthol) 106
  spectrophotometry using
    Arsenazo III 135
    diphenylcarbazone 194
    dithizone 123, 194
    8-hydroxyquinoline 99
Metal ion
  effect on complex stability 15–18
  classification according to
    Ahrland, Chatt and Davies 16–17
    Pearson 17
    Sidgwick 15–16
2-Methyl-8-hydroxyquinoline 50, 105
Mixed complexes 34, 49
  analytical selectivity of 47–49
  light absorption of 53
  rate of formation of 49
  solubility of 51
Molybdenum, determination by
  gravimetry using 8-hydroxyquinoline 104
  solvent extraction using
    acetylacetone 114
    8-hydroxyquinoline 104
  spectrophotometry using
    mercaptoacetic acid 195
    4-methyl-1,2-dimercaptobenzene 195
    thiomalic acid 195
Morin, analytical application of 111–112

Naphtholaldoximes 62
Neocuproine, analytical application of 81–82

SUBJECT INDEX

Nickel,
  determination by
    gravimetry using
      anthranilic acid 108
      dimethylglyoxime 83
      2,2'-dipyridyl + SCN^- 82
      1,10-phenanthroline + SCN^- 81
    polarography 76, 87–88
    precipitation from homogeneous solution 69
    solvent extraction using
      acetylacetone 113
      diethyldithiocarbamate 126
      dimethylglyoxime 85
      dithizone 120, 123
      α-furyldioxime 88
      8-hydroxyquinoline 100
      PAN (pyridylazonaphthol) 107
      TTA (thenoyltrifluoroacetone) 117
    spectrophotometry using
      α-benzyldioxime 196
      dimethylglyoxime 59–61, 196
      dimethylglyoxime + Br_2 61, 196
      Komplexon III 196
      β-mercaptopropionic acid 196
      sodium diethyldithiocarbamate 196
    volumetry using dimethylglyoxime 71–72, 83–85
  reaction with
    dimethylglyoxime 39, 59–61
    dithizone 49
Niobium, determination by
  gravimetry using morin 111
  solvent extraction using
    acetylacetone 114
    TTA (thenoyltrifluoroacetone) 117
  spectrophotometry using
    8-hydroxyquinoline 196
    pyrogallol 196
    sodium diethyldithiocarbamate 127
Nioxime (1,2-cyclohexanedionedioxime)
  analytical application of 61, 88
  stability constants of complexes 223
Nitrilotriacetate complexes, stability constants of 224
α-Nitro-β-naphthol 94–95
Nitrosonaphthols
  analytical application of 11, 57–58, 78, 92–93
  stability constants of its complexes 212
Nitroso R salt (1-nitroso-2-naphthol-3,6-disulfonic acid) 57

Organic solvents 232–235
Osmium, determination by
  spectrophotometry using
    PAN (pyridylazonaphthol) 107
    rubeanic acid 130
    thiourea 195

Oxidimetric measurements 65, 71–72, 109, 110–111, 131

Palladium,
  determination by
    gravimetry using
      dimethylglyoxime 83
      salicylaldoxime 89–90
    solvent extraction using
      acetylacetone 114
      diethyldithiocarbamate 128
      dimethylglyoxime 83, 197
      dithizone 120, 123
      α-furyldioxime 88
      8-hydroxyquinoline 100, 197
      PAN (pyridylazonaphthol) 107
    spectrophotometry using
      dimethylglyoxime 197
      2,2'-dipyridylglyoxime 197
      8-hydroxyquinoline 197
  reaction with
    p-dimethylaminobenzylidenerhodanine 11
    dimethylglyoxime 38, 51
PAN 1-(2-(pyridylazo)-2-naphthol), analytical application of 105–107
1,10-Phenanthroline
  analytical application of 58, 72, 80–81
  stability constants of its complexes 21, 225
Phenylarsonic acid, analytical application of 132–133
Picrolonic acid, analytical application of 95
Platinum, determination by spectrophotometry using
  diphenylthiosemicarbazide 197
  dithizone 123
Polarographic measurements 75–78, 87–88
Polyamine chelates, stability constants of 19
Polyamine complexes
  analytical application of 72
  chelate effect in 19
  stability constants of 19–20
Potassium
  determination by gravimetry using Kalignost 137
    spectrophotometry using
      dipicrylamine + Li_2CO_3 194
      sodium tetraphenylborate 194
    volumetry using Kalignost 137
  reaction with dipicrylamine 66
Precipitation from homogeneous solution 69–70
Pyridine-2-aldoxime, stability constants of its complexes 226
Pyrogallol, analytical application of 63–64

Quercetin, analytical application of 112
Quinaldic acid, analytical application of 109–110

Rare earth metals
  determination 117, 135
  extraction by TTA 117
  gas chromatographic separation 115
Reagents, solubility of 50, 68
Rhenium, determination by spectrophotometry with
  2,4-diphenylthiosemicarbazide 197
  α-furyldioxime 197
  thiourea 197
Rhodanine and derivatives 65
Rhodium, extraction with
  PAN (pyridylazonaphthol) 198
  TTA (thenoyltrifluoroacetone) 116
Rubeanic acid (dithiooxamide)
  analytical application of 129–130
  stability constants of its complexes 216
Ruthenium, determination by spectrophotometry using
  dithiooxamide (rubeanic acid) 130, 198
  1,10-phenanthroline 198
  thiourea 198

Salicylaldoxime
  analytical application of 62–63, 89–90
  stability constants of its complexes 62, 227
  structure of its complexes 89
Scandium
  determination by
    spectrophotometry using
      Arsenazo III 135, 136
      8-hydroxyquinoline 198
      sodium alizarin sulfonic acid 198
    extraction by
      acetylacetone 114
      TTA (thenoyltrifluoroacetone) 117
Selective reactions 13, 15
Selectivity 54
  factor (S. F.) 36–38
  index 38–39
  of dioximes 59
  of dithiocarbamates 125
  of ferron 56
  of hydroxyquinaldine 105
  of mandelic acid 65
  of neocuproin 81
  of nitroso R salt 57
  of 1,10-phenanthroline 80
  of quinaldic acid 109–110
  of salicylaldoxime 62, 89
Silver, determination by
  reaction with rhodanine derivatives 65–66
  solvent extraction using dithizone 120, 122
  spectrophotometry using
    p-diethylaminobenzylidenerhodanine 186
    dithizone 122, 186

Soft acids and bases 17
Solubility
  of complexes 49–51, 72
  of ligands 68
  of reagents 50, 68
Specific reaction 54
Spectrochemical series of ligands 44
Spectrophotometric measurements 72–75 185–204
Stability constants 19, 20, 21, 22, 23, 24, 25, 62, 207–232
  analytical importance of 13, 15, 23–25
Stereochemical changes, effect of, on complex stability 20–22, 23–25
Steric effect 20–22, 23–24
Strontium, determination by
  solvent extraction using
    8-hydroxyquinoline 100
    TTA (thenoyltrifluoroacetone) 117
  spectrophotometry using
    Arsenazo III 135
    Murexide 199
Substituent effect
  on complex formation 21, 56, 116, 129
  on solubility 50
Synergism in solvent extraction 74
Sulfosalicylate complexes, stability of 227

Tantalum, determination by spectrophotometry using
  phenylfluorone 199
  pyrogallol 199
Tartaric acid, stability constant of its complexes 228
Tellurium, determination by solvent extraction with diethyldithiocarbamate 127
Thallium(III)
  determination by
    gravimetry using
      Kalignost 137
      thionalide 131
    solvent extraction using
      diethyldithiocarbamate 127
      dithizone 123
      8-hydroxyquinoline 100
      rubeanic acid 130
      TTA (thenoyltrifluoroacetone) 117
    spectrophotometry using
      Brilliant Green 200
      Methyl Violet 200
      Rhodamine B 201
      thionalide 200
    volumetry using
      Kalignost 137
      thionalide 131
  extraction by
    acetylacetone 114
    TTA (thenoyltrifluoroacetone) 117–118

SUBJECT INDEX

Thenoyltrifluoroacetone (TTA)
  analytical application of 116-118
  extraction of its complexes 117-118
  stability constants of its complexes 229
Thermogravimetric investigations 205-206
  of dioxime chelates 83
Thioacetamide 132
Thioglycollic acid anilide 131
Thionalide (thioglycollic acid-β-aminonaphthalide)
  analytical application of 130-131
  stability constants of its complexes 229
Thiourea complexes, stability constants of 230
Thorium, determination by
  solvent extraction using
    acetylacetone 114
    8-hydroxyquinoline 100
    TTA (thenoyltrifluoroacetone) 117
  spectrophotometry using
    Arsenazo I 134
    Arsenazo III 135-136, 199
    8-hydroxyquinoline 100
    morin 199
    quercetin 199
    thoron 199
  volumetry using Kalignost 137
Tin, determination by
  solvent extraction using
    diethyldithiocarbamate 127
    dithizone 120
  spectrophotometry using
    Dithiol 199
    2,3,7-trihydroxy-9-p-nitrophenyl-6-fluorone 199
Tiron
  analytical application of 200
  stability constants of complexes 231
Titanium, determination by
  gravimetry using phenylarsonic acid 132
  solvent extraction using 8-hydroxyquinoline 100
  spectrophotometry using
    Arsenazo III 135
    ascorbic acid 200
    chromotropic acid 200
    hydrocinnamic acid 200
    8-hydroxyquinoline 100
    sulfosalicylic acid 200
    tiron 200
Trifluoroacetylacetone 115
Tungsten, determination by spectrophotometry using
  dithiol + $SnCl_2$ 202
  hydroquinone 202
  8-hydroxyquinoline 202

Uranium, determination by
  solvent extraction using
    acetylacetone 114
    8-hydroxyquinoline 101
    PAN (pyridylazonaphthol) 107
    TTA (thenoyltrifluoroacetone) 117, 118
  spectrophotometry using
    acetylacetone 201
    Arsenazo III 135, 136, 201
    dibenzoylmethane 201
    8-hydroxyquinoline 201
    mercaptoacetic acid 201
    PAN (pyridylazonaphthol) 201
    sodium diethyldithiocarbamate 202

Vanadium, determination by
  solvent extraction using
    acetylacetone 114
    diethyldithiocarbamate 127
    8-hydroxyquinoline 101
    PAN (pyridylazonaphthol) 107
  spectrophotometry using
    aniline 202
    8-hydroxyquinoline 202
    pyrocatechol 202

Yttrium, determination by
  solvent extraction using
    acetylacetone 114
    PAN (pyridylazonaphthol) 107
    TTA (thenoyltrifluoroacetone) 117
  spectrophotometry using Arsenazo III 135

Zinc, determination by
  gravimetry using
    anthranilic acid 108
    8-hydroxyquinoline 101
  polarography in the presence of cobalt 75-76
  solvent extraction using
    acetylacetone 114
    dithizone 120, 121, 123
    8-hydroxyquinoline 101
    PAN (pyridylazonaphthol) 107
  spectrometry using
    Arsenazo III 135
    dithizone 121, 123, 203
    8-hydroxyquinoline 203
    Zincon 203
  volumetry using
    anthranilic acid 109
    8-hydroxyquinoline 104
Zirconium,
  determination by
    gravimetry using
      arsanilic acid 133
      mandelic acid 110
      phenylarosnic acid 132-133
    solvent extraction using
      acetylacetone 114

Zirconium *(contiuned)*
  solvent extraction using
    8-hydroxyquinoline 101
    TTA (thenoyltrifluoroacetone) 118
  spectrophotometry using
    Arsenazo I 134, 203
    Arsenazo III 135, 136, 203
    chloranilic acid 204
    p-dimethylaminoazophenylarsonic acid 204
    morin 111, 204
    phenylfluorone 204
    Pyrocathechol Violet 204
    quercetin 204
    sodium alizarin sulphonic acid 203
  volumetry using mandelic acid 110
reaction with mandelic acid 65

OTHER TITLES IN THE SERIES IN ANALYTICAL CHEMISTRY

Vol. 1. Weisz—Microanalysis by the Ring Oven Technique.
Vol. 2. Crouthamel—Applied Gamma-ray Spectrometry.
Vol. 3. Vickery—The Analytical Chemistry of the Rare Earths.
Vol. 4. Headridge—Photometric Titrations.
Vol. 5. Busev—The Analytical Chemistry of Indium.
Vol. 6. Elvell and Gidley—Atomic Absorption Spectrophotometry.
Vol. 7. Erdey—Gravimetric Analysis. Parts 1–3.
Vol. 8. Critchfield—Organic Functional Group Analysis.
Vol. 9. Moses—Analytical Chemistry of the Actinide Elements.
Vol. 10. Ryabchikov and Gol'braikh—The Analytical Chemistry of Thorium.
Vol. 11. Cali—Trace Analysis for Semiconductor Materials.
Vol. 12. Zuman—Organic Polarographic Analysis.
Vol. 13. Rechnitz—Controlled-potential Analysis.
Vol. 14. Milner—Analysis of Petroleum for Trace Elements.
Vol. 15. Alimarin and Petrikova—Inorganic Ultramicroanalysis.
Vol. 16. Moshier—Analytical Chemistry of Niobium and Tantalum.
Vol. 17. Jeffery and Kipping—Gas Analysis by Gas Chromatography.
Vol. 18. Nielsen—Kinetics of Precipitation.
Vol. 19. Caley—Analysis of Ancient Metals.
Vol. 20. Moses—Nuclear Techniques in Analytical Chemistry.
Vol. 21. Pungor—Oscillometry and Conductometry.
Vol. 22. J. Zyka—Newer Redox Titrants.
Vol. 23. Moshier and Sievers—Gas Chromatography of Metal Chelates.
Vol. 24. Beamish—The Analytical Chemistry of the Noble Metals
Vol. 25. Yatsimirskii—Kinetic Methods of Analysis.
Vol. 26. Szabadváry—History of Analytical Chemistry.
Vol. 27. Young—The Analytical Chemistry of Cobalt.
Vol. 28. Lewis, Ott and Sine—The Analysis of Nickel.
Vol. 29. Braun and Tölgyessy—Radiometric Titrations.
Vol. 30. Ružička and Starý—Substoichiometry in Radiochemical Analysis.
Vol. 31. Crompton—Analysis of Organoaluminium and Organozinc Compounds.
Vol. 32. Schilt—Analytical Applications of 1,10-Phenanthroline and Related Compounds.
Vol. 33. Bark and Bark—Thermometric Titrimetry.
Vol. 34. Guilbault—Enzymatic Methods of Analysis.
Vol. 35. Wainerdi—Analytical Chemistry in Space.
Vol. 36. Jeffery—Chemical Methods of Rock Analysis.
Vol. 37. Weisz—Microanalysis by the Ring Oven Technique (2nd Edition—enlarged and revised).
Vol. 38. Rieman and Walton—Ion Exchange in Analytical Chemistry.
Vol. 39. Gorsuch—The Destruction of Organic Matter
Vol. 40. Mukherjee—Analytical Chemistry of Zirconium and Hafnium.
Vol. 41. Adams and Dams—Applied Gamma Ray Spectrometry (2nd Edition).
Vol. 42. Beckey—Ionization Mass Spectrometry.
Vol. 43. Lewis and Ott—Analytical Chemistry of Nickel.
Vol. 44. Silverman—Determination of Impurities in Nuclear Grade Sodium Metal.

Vol. 45. KUHNERT-BRANDSTATTER—Thermomicroscopy in the Analysis of Pharmaceuticals.
Vol. 46. CROMPTON—Chemical Analysis of Additives in Plastics.
Vol. 47. ELWELL and WOOD—Analytical Chemistry of Molybdenum and Tungsten.
Vol. 48. BEAMITH and VAN LOON—Recent Advances in the Analytical Chemistry of the Noble Metals.
Vol. 49. TÖLGYESSY, BRAUN and KRYS—Isotope Dilution Analysis.
Vol. 50. MAJUMDAR—N-Benzoylphenylhydroxylamine and its Analogues.
Vol. 51. BISHOP—Indicators.
Vol. 52. PŘIBIL—Analytical Applications of E.D.T.A. and Related Compounds.
Vol. 53. BAKER and BETTERIDGE—Photoelectron Spectroscopy–Chemical and Analytical Aspects.